"十四五"职业教育国家规划教材

城市轨道交通客运服务英语口语

（第3版）

程 逆 主 编
侯 顿　许迅安　副主编
经 纬 主 审

人民交通出版社

北京

内 容 提 要

本书是"十四五"职业教育国家规划教材、技工教育和职业培训"十四五"规划教材。随着我国城市轨道交通蓬勃发展，新业态不断显现，国际化专业人才需求不断增多，行业新知识、新技术、新工艺、新标准不断出现和完善。在此背景下，编写组对教材内容、框架进行了修订，以产教融合、校企"双元"育人为导向，以城市轨道交通行业为依托，融入课程思政；以学生为中心，运用丰富的教学栏目呈现教材内容；以数字化为载体助推一体化教学，适应"互联网+职业教育"发展需求。

根据职业标准和岗位需求，本书内容围绕：Green Transport、Directing Ways、Station Services、Smiling Services、Emergency五个部分，融入课程思政，序化出十个教学单元，实现课程内容与职业标准对接、岗位情景与英语语境对接、教学过程与服务过程对接。

本书可供职业院校城市轨道交通类专业教学使用，也可作为相关行业岗位培训用书，同时可供城市轨道交通从业人员学习参考。

本书配有丰富助学助教资源，请有需求的教师通过加入职教轨道教学研讨群（QQ群：129327355）获取。

图书在版编目（CIP）数据

城市轨道交通客运服务英语口语/程逆主编.—3版.—北京：人民交通出版社股份有限公司，2024.9

ISBN 978-7-114-19315-6

Ⅰ.①城… Ⅱ.①程… Ⅲ.①城市铁路—轨道交通—客运服务—英语—口语 Ⅳ.①U239.5

中国版本图书馆CIP数据核字（2024）第020477号

Chengshi Guidao Jiaotong Keyun Fuwu Yingyu Kouyu

书　　名：	城市轨道交通客运服务英语口语（第3版）
著　作　者：	程　逆
责任编辑：	钱　堃
责任校对：	赵媛媛　魏佳宁
责任印制：	刘高彤
出版发行：	人民交通出版社
地　　址：	（100011）北京市朝阳区安定门外外馆斜街3号
网　　址：	http://www.ccpcl.com.cn
销售电话：	（010）59757973
总 经 销：	人民交通出版社发行部
经　　销：	各地新华书店
印　　刷：	北京市密东印刷有限公司
开　　本：	787×1092　1/16
印　　张：	10
字　　数：	308千
版　　次：	2017年8月　第1版 2021年11月　第2版 2024年9月　第3版
印　　次：	2024年9月　第3版　第1次印刷
书　　号：	ISBN 978-7-114-19315-6
定　　价：	42.00元

（有印刷、装订质量问题的图书，由本社负责调换）

版 权 声 明

本教材自2017年出版第1版起,历经作者团队从创编到修订,至今第3版。其间,从相关调研到资料收集与整理加工,及至成书,倾注了作者大量心血。所形成的核心内容及内容序化与文字表述均为创作性成果。

人民交通出版社股份有限公司依法对本教材享有专有出版权,本书作者对书中原创性成果享有著作权(已在正文或参考文献中标注的有关引用部分除外)。任何未经许可的复制、传播和不当引用行为均违反《中华人民共和国著作权法》,其行为人将承担相应的法律责任。

特此声明。

第3版前言 PREFACE OF THE 3RD EDITION

《城市轨道交通客运服务英语口语（第3版）》入选"十四五"职业教育国家规划教材、技工教育和职业培训"十四五"规划教材。教材得到了众多院校的肯定和广泛使用，也收获了有价值的反馈意见。随着我国城市轨道交通蓬勃发展，新业态不断显现，国际化专业人才需求不断增多，行业新知识、新技术、新工艺、新标准不断出现和完善，充实和更新本教材的内容显得尤为重要。鉴于此，在《国务院关于印发国家职业教育改革实施方案的通知》（国发〔2019〕4号）"产教融合，'双元'育人"改革理念的指导下，结合《城市轨道交通服务员国家职业技能标准》和岗位要求，本教材编写组对教材内容、框架进行了修订，完成了第3版教材的编写。希望读者通过对本教材的学习，能够掌握城市轨道交通专业的理论知识和职业技能要求，提升英语听说水平，增强跨文化交流能力。

本教材主要具备以下特点：

一、以产教融合、校企"双元"育人为导向，顺应职教改革要求

本教材从技术技能人才成长规律和学生认知特点出发，通过校企合作、产教融合，邀请企业专家和优秀教师组成编写团队，围绕五个主题、十个单元，设计课程的整体框架。其中编写团队由武汉铁路职业技术学院程逆、侯顿、许迅安和天津三号线轨道交通运营有限公司贾瑞智组成，主审由天津三号线轨道交通运营有限公司经纬担任。本教材在强调职教特色的同时，体现了适用性、适宜性。

二、以城市轨道交通行业为依托，融入课程思政构建内容，提升职业素养

教材坚持立德树人的根本要求，渗透对学生综合素质的培养，结合《城市轨道交通服务员国家职业技能标准》、行业标准和岗位要求，构建教材内容。围绕Green Transport、Directing Ways、Station Services、Smiling Services、Emergency五个部分，融入绿色生态文明教育、融通中外、专业精神、职业精神、工匠精神、劳模精神和服务意识的思政主题，设计

出Introduction to Urban Rail Transit、Metro Guide、Timetable and Transfer、Tourist Attractions、Tickets and Fares、Checking Tickets、Station Facilities、Customer Service Center、Service Etiquette、Emergency Handling课程单元，涉及北京地铁、上海地铁、武汉地铁、广州地铁、青岛地铁、天津地铁、西安地铁和香港地铁等的相关内容以及国外城市轨道交通英语表达，实现课程内容与职业标准对接、岗位情景与英语语境对接、教学过程与服务过程对接，有助于培养具有国际视野、充满正能量、德智体美劳全面发展的城市轨道交通行业应用型人才。

三、以学生为中心设计丰富的教学栏目，增强教材吸引力，提升跨文化交流能力

与传统专业英语教材不同，本教材遵循学生的内在需求，每个单元设计由易到难、由浅入深，涵盖Situations（学习情境）、Goals（学习目标）、Groups（任务分组）、Knowledge contents（知识储备）、Task sheet（口语训练单）、Complementary reading（拓展阅读）、Relevant knowledge（知识链接）、Summary（总结）。在技能与语言的运用上，结合活页式Task Sheet（口语训练单）营造出真实工作情景，重点培养英语口语能力；图文并茂，语言训练形式多样，将抽象变形象、枯燥变生动，从英语的听、说、读、写四个方面重点培养学生的语言应用能力，突破英语学习瓶颈，提升口语交流水平。值得一提的是课程中设计了互动Role（职业角色），并贯穿始终，还原真实城市轨道交通工作环境、岗位任务和对客服务环节，可增强学生学习兴趣，提升职业体验感和认同感。

四、配套丰富信息化资源，适应"互联网+职业教育"发展需求

为方便学生的自主化学习和教师的信息化教学，本教材编写团队开发了包含课程标准、课件、微课和习题等在内的课程资源和不断完善的网络平台资源，将静态变动态、单向变双向，提升教材实施能力，构建线上线下良好教学生态。

在线课程

五、方便活页式装订，已印刷活页孔位置

为更好地贯彻执行《国务院关于印发国家职业教育改革实施方案的通知》（国发〔2019〕4号）中"倡导使用新型活页式、工作手册式教材并配套开发信息化资源"的理念，教材在"任务化"教学内容的基础上，在全书印刷了活页孔位置，教师和学生可根据自身需求，将教材拆分打孔后放入B5纸张9孔型标准活页夹或使用装订环，装订成活页式教材使用。

装订成活页式教材后，本教材可根据实际教学需求进行灵活调整，实

现"教材""学材"的融合及提升，并新增以下特点：

1. 方便"教材"的内容组合与动态更新

（1）可凸显教材内容的项目化、模块化、任务化设计，方便教学团队组织教学，可根据教学需求调整教学顺序；

（2）可根据不同使用对象、不同专业的教学要求，替换、添加、删减教学内容和教辅资料；

（3）可结合行业热点、最新时事、典型案例等，随时补充教学素材；

（4）可促进"岗课赛证"融通，将岗位职业技能、专业教学标准、技能大赛、"1+X"职业技能等级证书的内容灵活补充到教材中；

（5）可方便任务实训工单的收缴，评分后返给学生。

2. 方便"学材"的内容整理和灵活使用

（1）可随时添加学习笔记、学习心得到教材对应位置，方便复习；

（2）可灵活添加学习辅助资料，如参考资料、习题等；

（3）可根据上课内容携带对应页码，不用带整本书，简单方便；

（4）可根据自我学习进度随时调整学习顺序。

3. 方便教师和学生自由选择教材形式

（1）教材在"任务化"教学内容的基础上，在全书印刷了活页孔位置，教师和学生可根据自身需求，将教材拆分打孔后放入B5纸张9孔型标准活页夹或使用装订环，装订成活页式教材使用。

（2）教材是在胶订的基础上印刷了活页孔位置，可供不想拆分成活页式教材的使用者按照非活页式教材使用。

主　编
2024年7月

"岗课赛证" 融通

"岗"——国家职业技能标准相关工种/岗位技能要求在教材中的融入

国家职业技能标准城市轨道交通服务员		站务员										行车值班员				
		五级/初级工				四级/中级工				三级/高级工		五级/初级工			四级/中级工	
		行车组织与施工组织	客运与服务	票务运作	应急情况处理	行车组织与施工组织	客运与服务	票务运作	应急情况处理	票务运作	行车组织与施工组织	客运与服务	票务运作	应急情况处理	客运与服务	票务运作
第1部分	第1课	√	√	√	√	√	√	√	√	√	√	√	√	√	√	√
	第2课	√	√	√	√	√	√	√	√		√	√	√		√	√
第2部分	第3课	√	√	√	√	√	√	√		√		√	√		√	√
	第4课	√	√				√					√			√	
第3部分	第5课		√	√	√		√	√	√	√		√	√	√	√	√
	第6课		√	√	√		√	√	√		√	√	√		√	√
第4部分	第7课	√	√	√			√	√	√	√		√	√	√	√	√
	第8课		√	√	√		√	√	√	√	√	√	√		√	√
	第9课		√	√	√		√	√	√	√		√	√	√	√	√
第5部分	第10课	√	√	√	√		√	√	√	√	√	√	√	√	√	√

续上表

教学改革理念		"课"——教学改革在教材中的融入		
		教师活动	学生活动	教学意图
第1部分	第1课	1.组织教学。	1.起立致礼，考勤。	1.德育及行为教育，上课准备。
	第2课	2.复习旧课；课程整体框架、课程标准、课程学习要求和考核要求。	2.展示预习，讨论，回答问题。	2.帮助学生养成复习、预习的习惯。
第2部分	第3课		3.通过图片认识本课主要词汇，释意连线；口语讨论练习。	3.通过图片展示激发学生的学习兴趣，引导学生理解本章任务表达的内容，发现学习难点，口语讨论学生表达问题，针对问题调整教学策略。
	第4课	3.课程导入。		
第3部分	第5课	4.学习新课。	4.听力练习；跟读；不同情境对话的学习；词汇、常用句型；语法学习；完成知识点练习训练；思维导图阅读；拓展阅读。	4.通过听力训练，增强对话词汇和知识点的理解；通过跟读，纠正发音，掌握本章节知识点，对话演，强化理解；图片知识点，知识拓展；提升跨文化交流能力；知识拓展。
	第6课			
	第7课	5.课程小结。	5.学习自评，对课程学习的自我评价。	5.归纳总结。
第4部分	第8课			
	第9课	6.考核评价	6.课后完成作业及预习任务；	6.巩固提高，查缺补漏，增强英语学习自信心。
第5部分	第10课			

续上表

"赛"——技能大赛赛项要求在教材中的融入

职业技能大赛		城市轨道交通行车值班员职业技能大赛										城市轨道交通行车值班员（信号）职业技能大赛			城市轨道交通列车司机职业技能大赛					
		理论知识										理论知识			理论知识					
		职业道德		基本要求						相关知识										
					基础知识															
		职业道德基本知识	职业道德守则	行车安全、消防安全、用电安全、公共安全防范	列车运行图、运营时刻表基础知识	客运组织基础知识	设备设施故障、恶劣天气、突发大客流、涉及公共安全等特殊情况下的应急处置	突发事件应急处置和各级应急预案知识	票务管理的相关知识	公共区火灾应急处置	站厅火灾(A/B端)	社会责任与职业道德	城市轨道交通安全基础知识	城市轨道交通乘务管理知识	社会责任与职业道德	安全基础知识	相关法律法规知识	乘客服务知识	行车组织知识	安全规章制度
第1部分	第1课	√		√								√			√	√		√		√
	第2课	√		√		√						√			√	√		√		√
第2部分	第3课		√	√	√								√						√	
	第4课		√		√				√											
第3部分	第5课				√	√			√					√				√		
	第6课					√			√					√						
	第7课					√	√	√	√					√			√		√	√
第4部分	第8课	√		√			√	√				√				√		√		√
	第9课	√							√									√	√	√
第5部分	第10课			√		√	√	√						√		√	√	√	√	√

续上表

"1+X" 职业技能等级证书技能要求		"证" —— "1+X" 职业技能等级证书技能要求在教材中的融入								
		初级				中级			高级	
		行车组织及施工组织	客运服务	票务运作	应急情况处理	客运服务	票务运作	应急情况处理	客运服务	应急情况处理
第1部分	第1课	√	√	√	√	√	√	√	√	√
	第2课	√	√	√	√	√	√	√	√	√
第2部分	第3课	√	√	√		√	√		√	
	第4课		√	√		√	√		√	
第3部分	第5课	√	√	√	√	√	√	√	√	√
	第6课	√	√	√	√	√	√	√	√	√
	第7课		√	√	√	√	√	√	√	√
第4部分	第8课	√	√	√	√	√		√	√	√
	第9课	√	√	√		√			√	√
第5部分	第10课	√	√	√	√	√	√	√	√	√

听力材料二维码资源

序号	名称	二维码
1	Lesson 1 听力	
2	Lesson 2 听力	
3	Lesson 3 听力	
4	Lesson 4 听力	
5	Lesson 5 听力	
6	Lesson 6 听力	
7	Lesson 7 听力	
8	Lesson 8 听力	
9	Lesson 9 听力	
10	Lesson 10 听力	

目录 CONTENTS

Part One	Green Transport	第1部分	绿色出行	1
Lesson 1	Introduction to Urban Rail Transit	第1课	城市轨道交通简介	2
Lesson 2	Metro Guide	第2课	地铁简介	15
Part Two	Directing Ways	第2部分	指引服务	29
Lesson 3	Timetable and Transfer	第3课	时刻表与换乘	30
Lesson 4	Tourist Attractions	第4课	旅游景点	41
Part Three	Station Services	第3部分	站务服务	57
Lesson 5	Tickets and Fares	第5课	车票与票价	58
Lesson 6	Checking Tickets	第6课	检票	71
Lesson 7	Station Facilities	第7课	车站设备	84
Part Four	Smiling Services	第4部分	微笑服务	99
Lesson 8	Customer Service Center	第8课	乘客服务中心	100
Lesson 9	Service Etiquette	第9课	服务礼仪	113
Part Five	Emergency	第5部分	紧急情况	127
Lesson 10	Emergency Handling	第10课	应急处理	128
Appendix		附录		141
	Main Role Profile		主要人物角色介绍	142
References		参考文献		143

Part One
Green Transport

第1部分　绿色出行

Greener traffic, cleaner air.
交通更绿色，空气更清新。

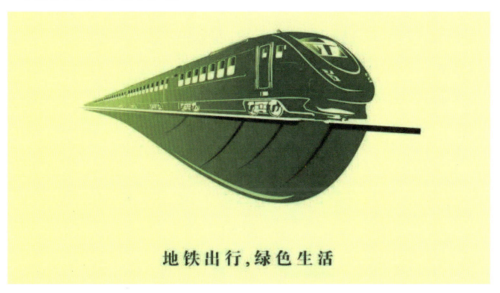

For a better living environment for the next generation, for the tomorrow of Mother Earth, let us advocate together: Metro travel, green life.

希望每一位地球居民，为了下一代更美好的生活环境，为了地球母亲的明天，让我们一起倡导：地铁出行，绿色生活!

Lesson 1　Introduction to Urban Rail Transit
第1课　城市轨道交通简介
Lesson 2　Metro Guide
第2课　地铁简介
Suggested hours: 6 class hours

Lesson 1　Introduction to Urban Rail Transit
第1课　城市轨道交通简介

📚 Overview

Situations (学习情境)

Goals (学习目标)

Groups (任务分组)

Knowledge contents (知识储备)

　　Lead-in (导入)

　　Listening (听力)

　　Conversations (情境对话)

Task sheet (口语训练单)

　　Objectives (训练目标)

　　Tasks (训练内容)

　　Self-evaluation (自我评价)

　　Grade evaluation (成绩评定)

　　Task reflection (任务反思)

Complementary reading (拓展阅读)

Relevant knowledge (知识链接)

Summary (总结)

📚 Situations

- A service agent is receiving a passenger at a transfer station.
- A passenger is asking for travel information.
- A service agent and a passenger are chatting on the Beijing Subway.

　　Q：假如你是一名站务员（service agent），你到车站接待外国客人，一起谈论旅行和北京的城市轨道交通，你会如何用英语开口表达呢？

2

Lesson 1　Introduction to Urban Rail Transit 城市轨道交通简介

Name:_____　Class:_____　Date:_____

Goals

- Introduce the development of Urban Rail Transit
- Master terminology
- Improve oral communication skills
- Improve professionalism and service awareness

Groups

Class（班级）		Group No.（小组编号）	
Leader（组长）		Facilitator（指导老师）	
Members（组员）		Roles（角色分工）	

Knowledge contents

Lead-in

1. Match the words and pictures

Monorail transit　　　　　Light rail transit　　　　　Metro

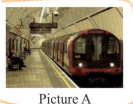
Picture A

Picture B

Picture C

2. Match the pictures and sentences

Picture A

(1) It is a term coined in the 1970s, during the re-emergence of streetcars. In general, it refers to streetcar with rapid transit-style features; it is named to distinguish it from heavy rail.

Picture B

(2) It is a single rail serving as a track for a wheeled or magnetically levitating vehicle; also, a vehicle traveling on such a track.

Picture C

(3) It usually runs in an urban area, and usually has high capacity and frequency, with large trains from other traffic. It is usually called the *underground* in BrE(British English) and the *subway* in AmE(American English).

Lesson 1　Introduction to Urban Rail Transit 城市轨道交通简介

Name:_____　　Class:_____　　Date:_____

3. Discuss and share

💡 **What is the most important reason for the development of urban rail transit? Please share your opinions with the others.**

Listening

1. Listen and tick the topic

London Underground ☐
Wuhan Metro ☐
New York Subway ☐

2. Listen and complete the blanks

| transportation | reduce | outside | crowded | 1863 |
| 30000 | | London | track | depend on |

When we talk about urban rail transit, there are some pictures coming to mind: such as Wuhan Metro, New York Subway and London Underground.

In some cities, the urban rail transit is so comprehensive and efficient that the majority of city residents travel without an automobile. Do you know which city has the first metro in the world? The answer is _____.The world's first metro was built in London in _____. The government was looking for a way to _____ traffic problems at the time. The poor areas of the city were so _____ with people that it was almost impossible for horse carriages to get through. On the other hand, the city officials were interested in trying to make it possible for workers to live _____ London, and travel easily to work each day.

If people had a cheap and convenient way that they could _____ to go to and from work, they would relocate their homes outside the city. This would help ease the pressure of too many people living in the poor parts of London. In this case, a convenient way of _____ appeared, the idea of the London Underground, the first subway system, was born.

A steam train pulled the cars along the fast underground _____ which was 6 kilometers (3.7 miles) long. About _____ people got on the subway the first day. Riders were treated to comfortable seats (standing up while the train was moving was not allowed), with pleasant decorations inside each of the cars.

3. Words and expressions

urban rail transit			城市轨道交通
comprehensive	[ˌkɒmprɪˈhensɪv]	adj.	全部的；所有的
city residents			城市居民
reduce traffic problems			减少交通问题

4

Lesson 1   Introduction to Urban Rail Transit 城市轨道交通简介

Name:_____   Class:_____   Date:_____

horse carriages			马车
get through			完成，度过，通过（考试）；及格
decoration	[ˌdekə'reɪʃən]	n.	装饰品；装饰
With its simple decoration, the main bedroom is a peaceful haven.			主卧室装饰简单，像一处宁静的港湾。
on the other hand (from another point of view)			（从）另一方面（来说）（用于引出相对照的另一观点）

4. Read after the recording

Conversations

1. Vocabulary preview

Public transport

The system of buses, trains, etc. provided by the government or by companies, which people use to travel from one place to another.

Most of us use public transport to get to work.

Traffic congestion

The state of being crowded and full of traffic.

Parking near the school causes traffic congestion.

Historical sites

Historical relics.

There are many famous scenic spots and historical sites in Beijing.

2. Dialogues

Scene A: A service agent is receiving a passenger at a transfer station

H-Helen (Service agent); M-Mr. Thomas (Passenger)

H: Excuse me, are you Mr. Thomas from New York Transportation Company?

M: Yes, I am.

H: My name is Helen Li, you can call me Helen. On behalf of Beijing Subway, I am here to meet you. Welcome to Beijing, Mr. Thomas!

M: Thank you, Helen. How do you do?

H: How do you do! How about your trip?

M: That's amazing! You know this is my first time to take China High-Speed Railway. I am still excited now.

H: Really? Just like flying?

M: Better than that! It takes only 5 hours from Shanghai to Beijing, traveling more than 1200 kilometers.

H: With a top speed of 300 kilometers per hour, our high-speed railway became the

Helen

Thomas

Lesson 1　Introduction to Urban Rail Transit 城市轨道交通简介

Name:_____　　Class:_____　　Date:_____

fastest train in the world. Passengers are deeply impressed by its high speed, stable operation and wonderful travel experience.

　　M: Yes. The food on train is delicious and the attendant brought them to me. Especially, the scenery outside the window is so beautiful. That's why I prefer to take the train.

　　H: How many pieces of luggage do you have?

　　M: Three.

　　H: Let me help with your luggage, Mr. Thomas.

　　M: Thank you, Helen. You are so kind.

　　H: Let's take Beijing Subway Line 4 to the hotel. This way, please.

　　M: OK, thank you!

Scene B: A passenger is asking for travel information

H-Helen (Service agent); M-Mr.Thomas (Passenger)

　　H: The meeting is drawing to a close. What are your plans for the next few days?

　　M: I would like to visit the scenic spots in Beijing. I have heard a lot about this city and now I can see it with my own eyes.

　　H: I hope you will enjoy your trip here. There are many scenic spots and historical sites in Beijing, such as the Great Wall, the Palace Museum, the Summer Palace, and so on.

Helen　　Thomas

　　M: I am really longing for that! But how to get there?

　　H: If you want to save energy to wander around the Palace Museum, you'd better take public transport. Bus and taxi are OK, but I recommend our urban rail transit.

　　M: As far as I know, urban rail transit consists of monorail transit, light rail transit, metro and so on, which one do you prefer?

　　H: Monorail transit is a vehicle traveling on a track and light rail transit refers to streetcar with rapid transit-style features. They are both typically less massive than that used for metro systems. So I think the metro is the most convenient.

　　M: Really? How to take it?

　　H: You can first download the Beijing Subway App on the Internet to check the routes and stations.

　　M: Oh, Helen, that's useful! Thank you for your advice.

Scene C: A service agent and a passenger are chatting on the Beijing Subway.

H-Helen (Service agent); M-Mr. Thomas (Passenger)

　　H: How about your visit in Beijing, Mr. Thomas?

　　M: Wonderful! Beijing is an ancient city with abandoned historical sites and natural landscapes. But what impresses me most is the Beijing subway.

　　H: Why?

　　M: With high service frequency, Beijing subway is the most convenient transport. Compared with the bus, taking the subway can avoid traffic congestion and save time; on the other hand, it is cheaper than the taxi. On the whole, I can have a point-to-point sightseeing.

　　H: Subway is really good, especially in rush hours.

Helen　　Thomas

Lesson 1   Introduction to Urban Rail Transit 城市轨道交通简介

Name:_____   Class:_____   Date:_____

M: Yeah, not only time-saving, but also reduce the discharge of harmful gas. Greener traffic, cleaner air!

H: Are you leaving tomorrow?

M: Yes. I will miss everything here: the Great Wall, the Palace Museum, Beijing roast duck, Peking Opera and Beijing subway!

3. Words and expressions

on behalf of			代表
high-speed rail			高速铁路
deeply	['diːpli]	adv.	很；非常；深刻地，强烈地
top	[tɒp]	n. adj.	顶，顶部 （位置、级别或程度）最高的；很好的
impress	[ɪm'pres]	v.	使敬仰；给……留下深刻的好印象
attendant	[ə'tendənt]	n.	服务员，侍者
stable	['steɪbl]	adj.	稳定的，稳固的
limousine	['lɪməziːn]	n.	礼宾车
save	[seɪv]	v.	保留，保存
useful	['juːsfʊl]	adj.	有用的；有益的；实用的
prefer ... to ...			相比……更喜欢……
the Great Wall			长城
the Palace Museum			故宫博物院
the Summer Palace			颐和园
Beijing roast duck			北京烤鸭
Peking Opera			京剧
on the other hand			（从）另一方面（来说）
This is my first time to...			这是我第一次……

4. Useful sentences and phrases

Welcome to Beijing.   欢迎来到北京。
That's amazing!   太棒了！
I can see it with my own eyes.   我能亲眼看到它。
The meeting is drawing to a close.   会议即将结束。
I am really longing for that!   我真的很期待！
Taking the subway can avoid traffic congestion and save time.   乘坐地铁可以避免交通拥挤，节省时间。

5. Grammar

point-to-point   点对点
one-to-one   一对一
door-to-door   门对门
luggage [ˈlʌɡɪdʒ] n.   行李(不可数名词)

Lesson 1   Introduction to Urban Rail Transit 城市轨道交通简介
Name:_____   Class:_____   Date:_____

Luggage is an uncountable noun. You can have a piece of luggage or some luggage but you cannot have *a luggage* or *some luggages*.

Luggage is the usual word in BrE, but *baggage* is also used, especially in the context of the bags and cases that passengers take on a flight. In AmE *baggage* is more usually used.

Both British and American speakers can refer to everything that travelers carry as *their bags*. American speakers can also call an individual suitcase *a bag*.

How many pieces of luggage do you have?

6. Exercises

Task 1: Fill in the blanks with the words from conversations that match the meanings of sentences

(1) If there is _____ in a place, the place is extremely crowded and blocked with traffic or people.

(2) _____ refers to streetcar with rapid transit-style features; it is named to distinguish it from heavy rail.

(3) _____ is a feeling of admiration for someone/something because you think they are particularly good.

(4) An _____ is someone whose job is to serve or help people on the train.

(5) _____ is a large expensive comfortable car.

(6) If something is _____, you can use it to do something or to help you in some way.

(7) _____ is a form of Chinese opera which combines music, vocal performance, mime, dance, and acrobatics.

(8) _____ is the suitcases or bags that you take with you when traveling.

Task 2: Fill in the blanks with the given words

top	limousine	on behalf of	prefer
stable	close	useful	deeply

(1) _____ Beijing Subway, I am here to meet you.
(2) He's one of the _____ singers in the world.
(3) My dear friends, today's metro trip is drawing to a _____. See you next time!
(4) Prices are _____ and the market is brisk.
(5) The advice you gave me was very _____.
(6) I will be _____ grateful to you for your kindness.
(7) He prefers traveling on the subway to riding in a _____.
(8) I _____ metro to monorail transit.

Task 3: Translate the following sentences into English

（1）中国的长城被誉为是世界的七大奇迹之一。

（2）我喜欢乘坐地铁远胜乘坐出租车。

（3）中国高铁真是太棒了！

Lesson 1　Introduction to Urban Rail Transit 城市轨道交通简介

Name:_____　Class:_____　Date:_____

（4）我太累啦，我真的很渴望暑假。

（5）他遇到交通堵塞，所以迟到了。

（6）请帮我照看一下行李。

 Task sheet

Objectives

Learn how to describe the metro / light rail transit/monorail transit.
Learn how to discuss the advantages of metro.

Oral task of grading details

Specific requirements for the academic quality of College English in higher Vocational Education.

Levels（水平分类）	Qualitative description（质量描述）
Level 1 (general requirements)〔水平一（一般要求）〕	Level 1-1: Can basically understand clear pronunciation, slow speech in daily life and urban rail transit passenger service posts related topics〔能基本听懂发音清晰、语速较慢的日常生活语篇和职场（城市轨道交通客运服务岗位）相关话题〕
	Level 1-2: Can basically understand the relevant English materials of urban rail transit passenger transport service, understand the main contents and obtain key information; understand the cultural connotation and identify the professional terms of urban rail transit passenger transport service（能基本读懂、看懂城市轨道交通客运服务的相关英语资料，理解主要内容，能获取关键信息；领会文化内涵，能识别城市轨道交通客运服务的专业术语）
	Level 1-3: Can communicate with others on familiar topics in daily life and urban rail transit passenger transport service; the expression is basically accurate and fluent, and can briefly introduce workplace culture and METRO culture（能在日常生活中和城市轨道交通客运服务工作中就比较熟悉的话题与他人进行语言交流，表达基本准确、流畅；能简单介绍职场文化和地铁企业文化）
	Level 1-4: Can briefly express their experiences, opinions and feelings; the sentences are basically correct and expressed clearly（能简要表达自己的经历、观点、感受；语句基本正确，表达清楚，格式恰当）
	Level 1-5: Can meet the basic communication needs on familiar with daily life and urban rail transit passenger transport service（能就日常生活和城市轨道交通客运服务工作中熟悉的话题，满足基本沟通需求）
	Level 1-6: Be able to make a clear learning plan; to obtain learning resources through online and offline channels under the guidance of teachers（能制订明确的学习计划；能在教师引导下通过线上线下多种渠道获取学习资源）

Lesson 1　Introduction to Urban Rail Transit 城市轨道交通简介

Name:_____　　Class:_____　　Date:_____

Task 1. Describe the pictures
Task 1-1: Metro

💡 **Way out**

It can be described from the following aspects: Urban rail/Definition/Characteristics/Development/History...

Key words: high service frequency/public transportation/metro station/metro systems/metro train...

💡 **Language notes**

urban rail transit　　城市轨道交通

采用专用轨道导向运行的城市公共客运交通系统，包括地铁、轻轨、单轨、有轨电车、磁浮、自动导向轨道、市域快速轨道系统。

passenger transport service　　客运服务

为使用城市轨道交通出行的乘客提供的服务。

专业术语来源：《城市轨道交通工程基本术语标准》（GB/T 50833—2012）
（*Standard for basic terminology of urban rail transit engineering*）

💡 **Outlines**

_____

_____

_____

Task 1-2: Light rail transit/Monorail transit

Lesson 1　Introduction to Urban Rail Transit 城市轨道交通简介

Name:_____　Class:_____　Date:_____

Key words: streetcar/frequent service/light trains/public transportation/single train

💡 **Outlines**

_____

_____

_____

Task 2. Practice your oral English and Role play

Task 2-1: Scene A

Work in pairs. Compared with the bus, talk about the advantages of the metro/subway.

💡 **Useful expressions/patterns**

I prefer the metro/subway to the bus for three reasons...

To my mind, the advantages/disadvantages are...

💡 **Outlines**

_____

_____

_____

Task 2-2: Scene B

Work in pairs. Compared with the taxi, talk about the advantages of metro/subway.

💡 **Useful expressions/patterns**

Compare with the taxi, the subway has several advantages...

I think.../don't think so. Because...

💡 **Outlines**

_____

_____

_____

Self-evaluation

Item	Excellent (90~100分)	Good (80~90分)	Average (60~80分)	Pass (60分)	Fail (<60分)
Be able to use phrases to make conversations					
Take part in pair work and role play					

Lesson 1　Introduction to Urban Rail Transit 城市轨道交通简介

Name:_____　　Class:_____　　Date:_____

Continued

Item	Excellent（90~100分）	Good（80~90分）	Average（60~80分）	Pass（60分）	Fail（<60分）
Improve oral communication skills					
Be able to finish the exercises independently					
Review and preview lessons consciously					

Grade evaluation

Item（项目）	No.（序号）	Criteria（标准）	Score（分数）		
			Self-evaluation（自评）	Peer（学生互评）	Facilitator（指导教师）
General Evaluation（一般评价）（25%）	1	Teamwork, well-assigned jobs（能进行团队合作，能做到分工良好）			
	2	Group discussions for the topic, plan for oral task materials（能进行小组讨论，能准备口语训练任务材料）			
	3	Well-prepared（准备充分）			
	4	Role playing well-performed（角色扮演表现良好）			
	5	Self-confidence（自信）			
Evaluation of Professional Competence（专业能力评价）（25%）	1	Pronunciation, articulation（发音清晰）			
	2	Accuracy, fluency（语言准确、流利）			
	3	Tone of voice, coherent（语调连贯）			
	4	Ability of cross-cultural communication（跨文化交流能力）			
	5	Presentation skills（口语技巧）			
Task-based Evaluation（基于任务的评估）（50%）	1	Describe the pictures（描述图片）			
	2	Practice your oral English and Role play（英语口语练习与角色扮演）			
Final Score（合计）		Self-evaluation Score × 20% + Peer Score × 30% + Facilitator Score × 50%（学生自评分数×20%+学生互评分数×30%+教师点评分数×50%）			
Facilitator's Comments（教师评价）					

Lesson 1   Introduction to Urban Rail Transit 城市轨道交通简介

Name:_____   Class:_____   Date:_____

Task reflection

FACT: What do you get?

FEELING: How do you feel?

FINDING: What do you find?

FUTURE: What shall you do next?

Complementary reading

China saw an increase in urban rail transit passenger trips

China saw an increase in passenger trips in its urban rail transit networks in April, 2023, official data showed.

The country's rail transit lines in urban areas registered 2.53 billion passenger trips in April, 2023, surging 95.8% year on year, data from the Ministry of Transport showed.

The figure represented an increase of 27.3% compared with the average monthly passenger trips in 2019, according to the data.

At the end of April, 2023 China had 292 urban rail transit lines in operation in 54 cities, with a total length of 9,652.6 km, according to the ministry.

(Source: Xinhua | 2023-05-08)

Translation:

(1) China saw an increase in passenger trips in its urban rail transit networks in April, 2023.

(2) The country's rail transit lines in urban areas registered 2.53 billion passenger trips in April, 2023.

Lesson 1   Introduction to Urban Rail Transit 城市轨道交通简介

Name:_____   Class:_____   Date:_____

(3) The figure represented an increase of 27.3% compared with the average monthly passenger trips in 2019.

(4) At the end of April, 2023 China had 292 urban rail transit lines in operation in 54 cities, with a total length of 9,652.6 km.

Relevant knowledge

Egypt's president inaugurates trial run of China-made LRT

CAIRO-Egypt's first electrified light rail transit (LRT) system jointly built by Chinese and Egyptian companies started its trial run on Sunday.

"The project was financed by China, adopted Chinese technology and equipment, and was jointly implemented by Chinese and Egyptian enterprises," Liao said, adding that all participants worked hard to keep the progress of construction while fighting against the COVID-19 pandemic.

The Chinese enterprises also provided training for young Egyptian technicians to support the construction and maintenance of the project, Liao added.

(Source: China Daily | 2022-07-05)

Summary

Complete the knowledge tree.

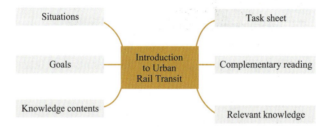

Lesson 2 Metro Guide
第2课 地铁简介

 **Overview**

Situations (学习情境)

Goals (学习目标)

Groups (任务分组)

Knowledge contents (知识储备)

    Lead-in (导入)

    Listening (听力)

    Conversations (情境对话)

Task sheet (口语训练单)

    Objectives (训练目标)

    Tasks (训练内容)

    Self-evaluation (自我评价)

    Grade evaluation (成绩评定)

    Task reflection (任务反思)

Complementary reading (拓展阅读)

Relevant knowledge (知识链接)

Summary (总结)

 **Situations**

- A passenger is asking for directions(1)
- Green transports make life better
- A passenger is asking for directions(2)

    Q：假如你是一名站务员（Service agent）或地铁公司实习生（Metro intern），有人问你如何去地铁站，你会怎样用英语开口表达呢？

Lesson 2　Metro Guide 地铁简介

Name:_____　　Class:_____　　Date:_____

Goals

- Provide help and information
- Master terminology
- Improve oral communication skills
- Improve professionalism and service awareness

Groups

Class（班级）		Group No.（小组编号）	
Leader（组长）		Facilitator（指导老师）	
Members（组员）		Roles（角色分工）	

Knowledge contents

Lead-in

1. Match the words and pictures

Traffic lights　　　　　　Metro map　　　　　　Metro station

Picture A

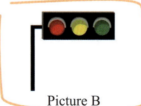
Picture B

Picture C

2. Match the pictures with the sentences

Picture A

(1) It contains a map of tracks, stops, and different lines that metro trains run on.

Picture B

(2) A place for the metro to stop, for the passengers to buy tickets and to wait for the train, with corresponding facilities.

Picture C

(3) They are sets of red, yellow, and green lights at the places where roads meet. They control the traffic by signaling red when vehicles have to stop and green when vehicles can go.

16

Lesson 2　Metro Guide 地铁简介

Name:_____　Class:_____　Date:_____

3. Discuss and share

💡 **Is there a subway in your hometown? If so, tell us about it. (For example: how many lines are there?)**

Listening

1. Listen and tick the topic

Providing help ☐
The way to Hilton Hotel ☐
How to get to the metro station ☐

2. Listen and complete the blanks

| between | down | good day | map | metro station |
| miss | train | far | left | traffic lights |

A: Excuse me, could you please tell me where I can find a _____?

B: Metro station? Certainly, the station is not _____ from here. You just go along this avenue until you can see some _____.

A: OK. Along this avenue ...

B: Yeah, that's right. And take a _____ turn.

A: OK. Turn left.

B: Then you will see the Central Bank and Hilton Hotel. The station is _____ the bank and hotel. You can't _____ it.

A: By the way, I want to go to Tiananmen Square, which _____ should I take?

B: Oh, when you go _____ to the metro station, just ask for a metro _____.

A: OK. I got it. That's very kind of you. Thanks a lot.

B: You're welcome. _____! Bye.

3. Words and expressions

metro station (subway station)			地铁站
go along			沿着……走
avenue	[ˈævənjuː]	n.	大街；林荫大道
turn left			左转
traffic lights			交通信号灯
Tiananmen Square			天安门广场

Lesson 2   Metro Guide 地铁简介

Name:_____   Class:_____   Date:_____

4. Read after the recording

Conversations

1. Vocabulary preview

Intersection

It is a place where roads or other lines meet or cross.
The cars slowed down as they approached the intersection.

Carbon dioxide

It is produced by animals and people when they breath out, and by chemical reactions. This gas can cause pollution and damage to the environment.
Taking the metro can reduce the emissions of carbon dioxide.

Museum

A depository for collecting and displaying objects having scientific or historical or artistic value to the public.
The art museum is usually closed on Mondays.

2. Dialogues

Scene A: A passenger is asking for directions(1)

A-Allen (Metro intern); P-Peter (Passenger)

P: Excuse me, I am sorry to bug you.

A: That's ok. What can I do for you, sir?

P: I am looking for a metro station. Can you show me where it is?

A: Of course, actually the nearest one is just nearby.

P: Oh, really?

A: Can you see that big mall over there?

P: Yes, I can see it.

A: The metro station is just at the back of the mall.

P: How can I get there?

A: You can go there on foot. It takes only a few minutes.

P: Thanks a lot.

A: Good day.

Allen

Peter

Scene B: Green transports make life better

H-Helen (Service agent); M-Mom(Helen's mom)

M: Finally, we are home!

H: Hey, mum. How was the trip?

M: It was great! The restaurant was fabulous, and the food was amazing. Your dad and I wish you were there with us.

H: I wish I could. But mum you know that I have to work.

M: Yes. I know, I know. You are busy. By the way, Wuxi now has the metro.

H: Really? Well, I don't know about that.

M: Only two lines are running so far, but they are still expanding.

H: Mum, from my point of view, the development of metro not only brings the

Helen

18

Lesson 2　Metro Guide 地铁简介

Name:_____　Class:_____　Date:_____

conveniences to our daily life, but also is good for the environment.

M: Absolutely. If everybody gets out by metro or takes other public transports, it will reduce the emissions of carbon dioxide and make our life better.

H: Tell me about it.

Scene C: A passenger is asking for directions(2)

H-Helen (Service agent); P-Peter (Passenger)

P: Excuse me, I think I am kind of lost. Could you please tell me the way to the metro station?

H: OK. You just need to walk along this road and then you will see an intersection. Turn right at that intersection, and you will see an advertising board. The metro station is at the left side of the board.

Helen　　Peter

P: Well …

H: Don't worry about it. It's easy to find it. Just remember to turn right at the intersection. You won't miss it.

P: OK, I got it. Thanks a lot. Bye.

H: You're welcome. Have a nice day!

3. Words and expressions

nearby	['nɪəbaɪ]	adv.	附近
route	[ruːt]	n.	线路
running	['rʌnɪŋ]	n.	运营
stretch	[stretʃ]	v.	延伸，扩建
bug	[bʌg]	n. vt.	昆虫，陷，瑕疵，细菌，窃听器 烦扰，打扰
metro station/subway station			地铁站
intersection	[ˌɪntə'sekʃən]	n.	十字路口
avenue	['ævənjuː]	n.	大街；林荫大道
fabulous	['fæbjʊləs]	adj.	极好的，棒的
station/stop		n.	站（请大家注意，在本书中"stop"的意思不是"停止"，而是"车站"）
originating station			始发站

4. Useful sentences and phrases

I am sorry to bug you. (I am sorry to bother you.)　　很抱歉打扰您。

I am kind of lost.　　我好像有点迷路了。

At the back of ...　　在……的后面

Compare with the bus, I prefer going out by metro.　　和公共汽车相比，我更喜欢乘地铁出门。

(*Compare with …, I prefer …* means that two choices offered, but you like one more than the other. For example: *Compare with pork, I prefer beef.*)

19

Lesson 2　Metro Guide 地铁简介

Name:_____　Class:_____　Date:_____

Finally, we are home!　　我们终于到家了！
I wish I could.　　我希望我能。
From my point of view ...　　在我看来……
(If you want to express your own thoughts, start by *from my point of view* is a very good way.)

5. Grammar

Could you tell me …?
Can you show me …?
宾语从句：
在句子中起宾语作用的从句叫作宾语从句。
宾语从句可以作及物动词、介词及形容词的宾语。
宾语从句一律用陈述句语序。
例如：*Could you tell me how I can go to the metro station?*
　　　Can you show me where it is?

6. Exercises

Task 1: Fill in the blanks with the words from conversations that match the meanings of sentences

(1) _____ is a place for the metro to stop, for the passengers to buy tickets and wait for the train, with corresponding facilities.

(2) _____ is a rapid transportation which runs underground.

(3) _____ is a place where you can draw money or deposit money.

(4) _____ is an awkward feeling when you have no idea about the directions.

(5) _____ is an important area of the avenue, where vehicles and people must follow the traffic lights.

Task 2: Fill in the blanks with the given words

| metro station　　back　　metro　　bank　　intersection
| traffic lights |

(1) _____ are placed at all major intersections.

(2) Something is wrong with my credit card. I need to go to the _____ to figure out the problem.

(3) I am sorry to bug you, but could you show me the way to the nearest _____?

(4) Be careful of the cars when you go through the _____.

(5) The post office is at the _____ of that building.

(6) The development of _____ not only brings the conveniences to our daily life, but also is good for the environment.

Task 3: Translate the following sentences into English

（1）与公交相比，我更喜欢搭乘地铁。

Lesson 2　Metro Guide 地铁简介

Name:_____　Class:_____　Date:_____

（2）不好意思，你能告诉我最近的地铁站在哪里吗？

（3）那个餐馆就在地铁站的后面。

（4）新的地铁线路将在明年开始运行。

（5）广告牌在十字路口的左边。

（6）地铁的确给我们的日常生活带来了便利。

Task sheet

Objectives

Learn how to describe the way to the metro station.
Learn how to offer some help.

Oral task of grading details

Specific requirements for the academic quality of College English in higher Vocational Education.

Levels（水平分类）	Qualitative description（质量描述）
Level 1 (general requirements)〔水平一（一般要求）〕	Level 1-1: Can basically understand clear pronunciation, slow speech in daily life and urban rail transit passenger service posts related topics〔能基本听懂发音清晰、语速较慢的日常生活语篇和职场（城市轨道交通客运服务岗位）相关话题〕
	Level 1-2: Can basically understand the relevant English materials of urban rail transit passenger transport service, understand the main contents and obtain key information; understand the cultural connotation and identify the professional terms of urban rail transit passenger transport service（能基本读懂、看懂城市轨道交通客运服务的相关英语资料，理解主要内容，能获取关键信息；领会文化内涵，能识别城市轨道交通客运服务的专业术语）
	Level 1-3: Can communicate with others on familiar topics in daily life and urban rail transit passenger transport service; the expression is basically accurate and fluent, and can briefly introduce workplace culture and METRO culture（能在日常生活中和城市轨道交通客运服务工作中就比较熟悉的话题与他人进行语言交流，表达基本准确、流畅；能简单介绍职场文化和地铁企业文化）
	Level 1-4: Can briefly express their experiences, opinions and feelings; the sentences are basically correct and expressed clearly（能简要表达自己的经历、观点、感受；语句基本正确，表达清楚，格式恰当）
	Level 1-5: Can meet the basic communication needs on familiar with daily life and urban rail transit passenger transport service（能就日常生活和城市轨道交通客运服务工作中熟悉的话题，满足基本沟通需求）
	Level 1-6: Be able to make a clear learning plan; to obtain learning resources through online and offline channels under the guidance of teachers（能制订明确的学习计划；能在教师引导下通过线上线下多种渠道获取学习资源）

Lesson 2　Metro Guide 地铁简介

Name:_____　Class:_____　Date:_____

Task 1. Describe the pictures
Task: Describe the way to the metro station

💡 **Way out**

Where is the metro station? How to get there?

💡 **Language notes**

trip　　出行
从出发地到目的地的交通行为。
trip volume　　出行量
单位时间内，居民出行的总人次数。
专业术语来源：《城市轨道交通工程基本术语标准》（GB/T 50833—2012）
（*Standard for basic terminology of urban rail transit engineering*）

💡 **Outlines**

_____

_____

_____

Task 2. Practice your oral English and Role play
Task 2-1: Scene A

Work in pairs. Suppose that you can't find the metro station and you are asking someone for help

💡 **Useful expressions/patterns**

Excuse me, I am new here. Could you please tell me...
Offer help. / Ok. The metro station is...

Lesson 2　Metro Guide 地铁简介

Name:_____　　Class:_____　　Date:_____

💡 **Outlines**

_____

_____

_____

Task 2-2: Scene B

Work in pairs. Suppose that someone who just got off a plane and couldn't find the metro station and he/she is asking you for a hand

💡 **Useful expressions/patterns**

Excuse me, I am just off the plane and I am looking for...

Well ...

💡 **Outlines**

_____

_____

_____

Self-evaluation

Item	Excellent (90~100分)	Good (80~90分)	Average (60~80分)	Pass (60分)	Fail (<60分)
Be able to use phrases to make conversations					
Take part in pair work and role play					
Improve oral communication skills					
Be able to finish the exercises independently					
Review and preview lessons consciously					

23

Lesson 2　Metro Guide 地铁简介

Name:_____　Class:_____　Date:_____

Grade evaluation

Item（项目）	No.（序号）	Criteria（标准）	Score（分数）		
			Self-evaluation（自评）	Peer（学生互评）	Facilitator（指导教师）
General Evaluation（一般评价）（25%）	1	Teamwork, well-assigned jobs（能进行团队合作，能做到分工良好）			
	2	Group discussions for the topic, plan for oral task materials（能进行小组讨论，能准备口语训练任务材料）			
	3	Well-prepared（准备充分）			
	4	Role playing well-performed（角色扮演表现良好）			
	5	Self-confidence（自信）			
Evaluation of Professional Competence（专业能力评价）（25%）	1	Pronunciation, articulation（发音清晰）			
	2	Accuracy, fluency（语言准确、流利）			
	3	Tone of voice, coherent（语调连贯）			
	4	Ability of cross-cultural communication（跨文化交流能力）			
	5	Presentation skills（口语技巧）			
Task-based Evaluation（基于任务的评估）（50%）	1	Describe the pictures（描述图片）			
	2	Practice your oral English and Role play（英语口语练习与角色扮演）			
Final Score（合计）		Self-evaluation Score × 20% + Peer Score × 30% + Facilitator Score × 50%（学生自评分数×20%+学生互评分数×30%+教师点评分数×50%）			
Facilitator's Comments（教师评价）					

Task reflection

FACT: What do you get?

FEELING: How do you feel?

FINDING: What do you find?

FUTURE: What shall you do next?

Lesson 2 Metro Guide 地铁简介

Name:_____ Class:_____ Date:_____

Complementary reading

Metro Survey

Fill in the form with the latest information

 Overview

Locale （区域）	Beijing	Daily ridership （日客流量）	
Transit type （运输类型）	Metro, light rail transit, tram, automated guideway transit	Annual ridership （年客流量）	
Number of lines （线路数量）		Began operation （开始运营时间）	1st October, 1969
Number of stations （车站数量）		Operator （运营企业）	Beijing MTR Corp. Ltd.

 Overview

Locale （区域）		Daily ridership （日客流量）	
Transit type （运输类型）		Annual ridership （年客流量）	
Number of lines （线路数量）		Began operation （开始运营时间）	28th May, 1993
Number of stations （车站数量）		Operator （运营公司）	Shanghai Shentong Metro Group

 Overview

Locale （区域）		Daily ridership （日客流量）	
Transit type （运输类型）		Annual ridership （年客流量）	
Number of lines （线路数量）		Began operation （开始运营时间）	
Number of stations （车站数量）		Operator （运营企业）	

Lesson 2　Metro Guide 地铁简介

Name:_____　Class:_____　Date:_____

 Overview

	Locale （区域）		Daily ridership （日客流量）	
	Transit type （运输类型）		Annual ridership （年客流量）	
	Number of lines （线路数量）		Began operation （开始运营时间）	
	Number of stations （车站数量）		Operator （运营企业）	

 Overview

	Locale （区域）		Daily ridership （日客流量）	
	Transit type （运输类型）		Annual ridership （年客流量）	
	Number of lines （线路数量）		Began operation （开始运营时间）	
	Number of stations （车站数量）		Operator （运营企业）	

Translation:

(1) Overview

(2) Transit type

(3) Rapid transit

(4) Number of lines

(5) Number of stations

(6) Daily ridership

Lesson 2   Metro Guide 地铁简介

Name:_____   Class:_____   Date:_____

 Relevant knowledge

Spring Festival embraces smart, green travel rush

During the rest of the travel rush season, the authorities should continue to promote green transportation, and take energy-saving and emission-reduction measures, in order to meet China's climate goals of realizing carbon emission peak before 2030 and achieve carbon neutrality before 2060. To do so, governments at all levels should more vigorously publicize green travel through railways and new energy vehicles, and promote large-capacity green transport modes. They should also improve their service by, for example, building more charging stations and parking lots for new energy vehicles.

(Source: China Daily | 2022-02-15)

 **Summary**

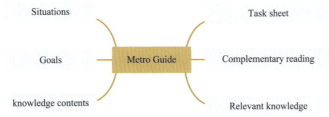

Part Two
Directing ways

第2部分 指引服务

A ticket will take you around the world.
一张票带你游遍世界。

It is better to travel thousands of miles than to thousands of books. The development of urban rail transit not only shorten the distance of the world, but also broaden our horizons.

"读万卷书,不如行万里路"。城市轨道交通的发展,不仅缩短了世界的距离,而且拓宽了我们的视野。

Lesson 3　Timetable and Transfer
第3课　时刻表与换乘
Lesson 4　Tourist Attractions
第4课　旅游景点
Suggested hours: 6 class hours

Lesson 3　Timetable and Transfer
第3课　时刻表与换乘

Overview

Situations (学习情境)

Goals (学习目标)

Groups (任务分组)

Knowledge contents (知识储备)

 Lead-in (导入)

 Listening (听力)

 Conversations (情境对话)

Task sheet (口语训练单)

 Objectives (训练目标)

 Tasks (训练内容)

 Self-evaluation (自我评价)

 Grade evaluation (成绩评定)

 Task reflection (任务反思)

Complementary reading (拓展阅读)

Relevant knowledge (知识链接)

Summary (总结)

Situations

- A student is asking for a transfer station
- A passenger is getting information of timetable
- Metro to the airport

　Q：假如你是一名站务员（service agent），你会如何为乘客提供换乘和地铁时刻表的信息？如何用英语开口表达呢？

Lesson 3   Timetable and Transfer 时刻表与换乘

Name:_____   Class:_____   Date:_____

Goals

- Introduce the timetable and transfer
- Master terminology
- Improve oral communication skills
- Improve professionalism and service awareness

Groups

Class（班级）		Group No.（小组编号）	
Leader（组长）		Facilitator（指导老师）	
Members（组员）		Roles（角色分工）	

Knowledge contents

Lead-in

1. Match the words and pictures

The Forbidden City   Timetable   Line

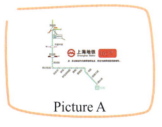
Picture A

Picture B

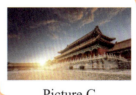
Picture C

2. Match the pictures and sentences

Picture A

(1) It is an imperial palace complex of the Ming and Qing dynasties in Beijing, China.

Picture B

(2) It is the direction that the train is moving.

Picture C

(3) It is a list of the times when trains are supposed to arrive at or leave from a particular place.

Lesson 3　Timetable and Transfer 时刻表与换乘

Name:_____　Class:_____　Date:_____

3. Discuss and share

💡 Have you ever help others to transfer lines? If you do, share it with us.

Listening

1. Listen and tick the topic

Inquiring about metro line ☐
Timetable ☐
Travel in Hong Kong ☐

2. Listen and complete the blanks

| take | line | wondering | useful | charge | fare | fun |

A: Excuse me, I am here for travel, so I was _____ if this metro goes to the Forbidden City?

B: Well, let me think... We are now in Dongdan（东单）. You can _____ Subway Line 1 to the Forbidden City.

A: Is this a direct _____?

B: Yes, you don't need to change another train.

A: Okay. How much is the _____?

B: Let me see, the _____ is 3 yuan.

A: By the way, can you tell me the way to the Forbidden City after getting out of the station?

B: Getting out of the northwest exit, you can see Tiananmen Square and the Forbidden City is just in front of you.

A: Okay. That is _____. Thanks for your time!

B: Wish you have _____ here!

3. Words and expressions

Excuse me			不好意思打扰一下
line	[laɪn]	n.	路线，轨道线路
Wish you have fun here!			希望你们玩得开心！

4. Read after the recording

Conversations

1. Vocabulary preview

Transfer

To change to another metro line or a different vehicle during a journey.
You just need to transfer to Wuhan Metro Line 1.

Lesson 3　Timetable and Transfer 时刻表与换乘

Name:_____　Class:_____　Date:_____

Stop

A place where a train stops regularly for passengers to get on or off.

I get off at the next stop.

Direction

The general position a train moves towards.

I am afraid you are in the opposite direction.

2. Dialogues

Scene A: A student is asking for a transfer station

H-Helen (Service agent); S-Student

Helen

S: Excuse me, how many stops are there to Sports Center (体育中心)?

H: Well … I'm afraid you are in the wrong direction.

S: What? Oh, My God! Sorry … I have just got here, so I am not quite familiar with this place.

H: That's okay. You just need to transfer to Line 1. There's a transfer station in Sports West Street (体育西路) and the next stop is Sports West Street, Sports Center (体育中心)is not far from that station.

S: And I don't need to buy a new ticket?

H: You don't.

S: Thanks for your kindness!

H: You're welcome! You will find it.

Scene B: A passenger is getting information of timetable

H-Helen (Service agent); P-Peter (Passenger)

Helen　Peter

P: I am sorry to bother you, I was wondering if this line goes to Cultural Park (文化公园)?

H: Yes, it is.

P: How long should I wait?

H: Well … A train has just passed, but you don't need to wait for a long time. Since it's rush hour, so each one arrives within 3 minutes.

P: That's fast! Thanks a lot!

H: Good day!

Scene C: Metro to the airport

A-Allen (Metro intern); B-His brother

A: Are you leaving tomorrow morning?

B: Yeah! I need to catch an early flight.

A: Oh. What a shame! I have an important meeting tomorrow morning, so I won't be able to drive you to the airport.

B: Hey, dude, that's okay. I can take the metro to the airport. Just tell me which line goes to the airport directly. I don't want any transfer.

Allen

A: Line 11. And make sure you are in the right direction. Okay?

B: And my flight is at 9 o'clock. I think I need to take the first one. I don't want to miss it.

A: The first one is at 6:30 a.m. And don't worry. You won't miss your flight.

B: I got it. Come on … Buddy, give me a hug! Thanks for your company these days. We'll meet again and I will miss you!

Lesson 3　Timetable and Transfer 时刻表与换乘

Name:＿＿＿＿＿　Class:＿＿＿＿＿　Date:＿＿＿＿＿

A: Alright. Good night! Sleep tight!
B: Nightly night!

3. Words and expressions

stop	[stɒp]	v.	停止，中断
		n.	停车站
quite	[kwaɪt]	adv.	相当
familiar	[fə'mɪljə(r)]	adj.	熟悉的
transfer	[træns'fɜː(r)]	v.	转移，转乘
dude	[duːd; djuːd]	n.	（俚语）哥们，小子
transfer station			换乘站
rush hour			高峰期

4. Useful sentences and phrases

How many stops are there ...?　　有多少站？
I'm afraid ...　　我恐怕……
Thanks for your kindness.　　谢谢你的好意。
What a shame! (What a pity!)　　太可惜！
Sleep tight. Nightly night. (Good night.)　　睡个好觉。
Which line goes to ...?　　哪一条线到……？
You are in the wrong direction.　　你弄错方向了。

5. Grammar

I was wondering if ...
委婉表达的用法：
此句型表示委婉的请求，不涉及时态问题。
例如：*I was wondering if I could borrow your phone.*
I was wondering if you could tell me how to fill out the form.

6. Exercises

Task 1: Fill in the blanks with the words from conversations that match the meanings of sentences

(1) ＿＿＿＿＿ is a station where you can transfer trains.
(2) ＿＿＿＿＿ is a line leading to a place or point.
(3) ＿＿＿＿＿ is a huge construction where you can do all kinds of sports.
(4) ＿＿＿＿＿ is a place where airplanes can take off.
(5) ＿＿＿＿＿ is a kind of transportation which can fly in the sky.

Task 2: Fill in blanks with the given words

| familiar | dude | direction | quite | transfer |

Lesson 3　Timetable and Transfer 时刻表与换乘

Name:_____　Class:_____　Date:_____

(1) Hey _____! I haven't seen you for few days. Where have you been?
(2) Can I _____ a train at this station?
(3) I'm new here, so I am not _____ with this place.
(4) The traffic is _____ busy in the rush hour!
(5) Sorry, you are in the wrong _____.

Task 3: Translate the following sentences into English

（1）还有多少站到学校？

（2）下一站是个换乘站。

（3）你弄错方向了。（地铁坐错了方向。）

（4）高峰期交通非常拥挤。

（5）哪一条线路到机场？

Task sheet

Objectives

Learn how to describe the timetable of metro.
Learn how to give directions.

Oral task of grading details

Specific requirements for the academic quality of College English in higher Vocational Education.

Levels （水平分类）	Qualitative description （质量描述）
Level 1 (general requirements) ［水平一（一般要求）］	Level 1-1: Can basically understand clear pronunciation, slow speech in daily life and urban rail transit passenger service posts related topics［能基本听懂发音清晰、语速较慢的日常生活语篇和职场（城市轨道交通客运服务岗位）相关话题］
	Level 1-2: Can basically understand the relevant English materials of urban rail transit passenger transport service, understand the main contents and obtain key information; understand the cultural connotation and identify the professional terms of urban rail transit passenger transport service（能基本读懂、看懂城市轨道交通客运服务的相关英语资料，理解主要内容，能获取关键信息；领会文化内涵，能识别城市轨道交通客运服务的专业术语）
	Level 1-3: Can communicate with others on familiar topics in daily life and urban rail transit passenger transport service; the expression is basically accurate and fluent, and can briefly introduce workplace culture and METRO culture（能在日常生活中和城市轨道交通客运服务工作中就比较熟悉的话题与他人进行语言交流，表达基本准确、流畅；能简单介绍职场文化和地铁企业文化）
	Level 1-4: Can briefly express their experiences, opinions and feelings; the sentences are basically correct and expressed clearly（能简要表达自己的经历、观点、感受；语句基本正确，表达清楚，格式恰当）

Lesson 3   Timetable and Transfer 时刻表与换乘

Name:_____   Class:_____   Date:_____

Continued

Levels（水平分类）	Qualitative description（质量描述）
Level 1 (general requirements) ［水平一（一般要求）］	Level 1-5: Can meet the basic communication needs on familiar with daily life and urban rail transit passenger transport service（能就日常生活和城市轨道交通客运服务工作中熟悉的话题，满足基本沟通需求）
	Level 1-6: Be able to make a clear learning plan; to obtain learning resources through online and offline channels under the guidance of teachers（能制订明确的学习计划；能在教师引导下通过线上线下多种渠道获取学习资源）

Tasks

Task 1. Describe the pictures

Lesson 3   Timetable and Transfer 时刻表与换乘

Name:_____   Class:_____   Date:_____

Suppose you were at CAOZHUANG（曹庄）, how do you get to NANLOU（南楼）?
Suppose you were at line 3, how to transfer line 2?
Suppose you were at FUXINGMEN（复兴门）, how do you get to ERDAOQIAO（二道桥）?
Suppose you were at XIZHAN（西站）, how do you get to DAXUECHENG（大学城）?

💡 **Language notes**

riding time/ ride time   乘行时间
在一次乘行中，乘客从上车到下车所花费的时间。
transfer   换乘
乘客在出行过程中转 换车次、线路、交通方式的行为。
departure time of first train   首班列车时间
首班列车驶离某车站的时刻。
departure time of last train   末班列车时间
末班列车驶离某车站的时刻。
专业术语来源：《城市轨道交通工程基本术语标准》（GB/T 50833-2012）
（*Standard for basic terminology of urban rail transit engineering*）

💡 **Outlines**

_____
_____
_____

Task 2. Practice your oral English and Role play

Task 2-1: Scene A

Work in pairs, suppose someone were asking you something about transferring to the metro

💡 **Useful expressions/patterns**

Excuse me, is this the transfer station? I want to …
Well …

💡 **Outlines**

_____
_____
_____

Task 2-2: Scene B

Work in pairs, suppose you were in the wrong direction and asked someone for help

💡 **Useful expressions/patterns**

How many stops are there to…?
What? You've just missed that station…

Lesson 3　Timetable and Transfer 时刻表与换乘

Name:_____　Class:_____　Date:_____

💡 Outlines

Self-evaluation

Item	Excellent（90~100分）	Good（80~90分）	Average（60~80分）	Pass（60分）	Fail（<60分）
Be able to use phrases to make conversations					
Take part in pair work and role play					
Improve oral communication skills					
Be able to finish the exercises independently					
Review and preview lessons consciously					

Grade evaluation

Item（项目）	No.（序号）	Criteria（标准）	Score（分数）		
			Self-evaluation（自评）	Peer（学生互评）	Facilitator（指导教师）
General Evaluation（一般评价）（25%）	1	Teamwork, well-assigned jobs（能进行团队合作，能做到分工良好）			
	2	Group discussions for the topic, plan for oral task materials（能进行小组讨论，能准备口语训练任务材料）			
	3	Well-prepared（准备充分）			
	4	Role playing well-performed（角色扮演表现良好）			
	5	Self-confidence（自信）			
Evaluation of Professional Competence（专业能力评价）（25%）	1	Pronunciation, articulation（发音清晰）			
	2	Accuracy, fluency（语言准确、流利）			
	3	Tone of voice, coherent（语调连贯）			

Lesson 3　Timetable and Transfer 时刻表与换乘

Name:_____　Class:_____　Date:_____

Continued

Item（项目）	No.（序号）	Criteria（标准）	Score（分数）		
			Self-evaluation（自评）	Peer（学生互评）	Facilitator（指导教师）
Evaluation of Professional Competence（专业能力评价）（25%）	4	Ability of cross-cultural communication（跨文化交流能力）			
	5	Presentation skills（口语技巧）			
Task-based Evaluation（基于任务的评估）（50%）	1	Describe the pictures（描述图片）			
	2	Practice your oral English and Role play（英语口语练习与角色扮演）			
Final Score（合计）		Self-evaluation Score × 20% + Peer Score × 30% + Facilitator Score × 50%（学生自评分数 × 20%+学生互评分数 × 30%+教师点评分数 × 50%）			
Facilitator's Comments（教师评价）					

Task reflection

FACT: What do you get?

FEELING: How do you feel?

FINDING: What do you find?

FUTURE: What shall you do next?

Complementary reading

General information about transfers with the 3-hour pass

- The new 3-hour pass will replace the 3-hour one-way transfer.
- The 3-hour pass will allow riders to take unlimited trips at the same service level on bus or train within 3 hours from purchase or issue. Passengers must pay an upgrade if they use a higher level of service.

Lesson 3　Timetable and Transfer 时刻表与换乘

Name:_____　Class:_____　Date:_____

- 3-hour passes issued from the on-bus printers are marked with the expiration time (3 hours from time of issue), and are valid until the time printed on the pass. Factors may extend the expiration time such as service delays, detours, and inclement weather.
- 3-hour passes are not valid fare on Access-a-Ride.
- A validated rail ticket may serve as a transfer when continuing a trip by bus. If passengers wish to use a ticket from a 10-Ride Ticket Book or a Free Ride Coupon on rail, the ticket or coupon must be validated at the ticket machine at the rail station.

Translation:

(1) The new 3-hour pass will replace the 3-hour one-way transfer.

(2) Passengers must pay an upgrade if they use a higher level of service.

(3) 3-hour passes are not valid fare on Access-a-Ride.

(4) A validated rail ticket may serve as a transfer when continuing a trip by bus.

(5) The ticket or coupon must be validated at the ticket machine at the rail station.

Relevant knowledge

Elements of Chinese culture in a Moscow subway station

Chinese technology, Chinese manufacturing, Chinese culture, and Chinese wisdom have all accompanied the "Belt and Road" initiative to the world.

Summary

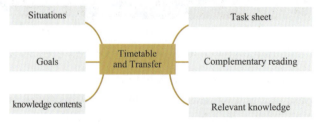

Lesson 4　Tourist Attractions
第4课　旅游景点

Overview

Situations (学习情境)

Goals (学习目标)

Groups (任务分组)

Knowledge contents (知识储备)

    Lead-in (导入)

    Listening (听力)

    Conversations (情境对话)

Task sheet (口语训练单)

    Objectives (训练目标)

    Tasks (训练内容)

    Self-evaluation (自我评价)

    Grade evaluation (成绩评定)

    Task reflection (任务反思)

Complementary reading (拓展阅读)

Relevant knowledge (知识链接)

Summary (总结)

Situations

- A passenger is asking about tourist attractions (1)
- A passenger is asking about tourist attractions (2)
- A passenger is requiring a way to a local attraction
- A service agent and her friend are talking about the itinerary

    Q：假如你是一名站务员（Service agent），你会如何向乘客提供旅游导览服务呢？如

Lesson 4　Tourist Attractions 旅游景点

Name:_____ Class:_____ Date:_____

何用英语开口表达呢？

🎓 Goals

- Describe a popular scenic spot
- Master terminology
- Improve oral communication skills
- Improve professionalism and service awareness

🎓 Groups

Class（班级）		Group No.（小组编号）	
Leader（组长）		Facilitator（指导老师）	
Members（组员）		Roles（角色分工）	

🎓 Knowledge contents

Lead-in

1. Match the words and pictures

Terracotta army　　　　Xi'an City Wall　　　　Goose

Picture A

Picture B

Picture C

2. Match the pictures and sentences

Picture A

(1) It is a large bird that has a long neck and webbed feet.

Picture B

(2) It refers to the thousands of life-size clay models of soldiers, horses, and chariots which were deposited around the grand mausoleum of the First Qin Emperor.

Picture C

(3) It surrounds the center of Xi'an and is one of the oldest, largest, and most complete city walls in China.

Lesson 4　Tourist Attractions 旅游景点

Name:_____　Class:_____　Date:_____

3. Discuss and share

💡 Do you like traveling? With who? Where have you been? How about the local food and attractions?

Listening

1. Listen and tick the topic

They are talking about the weather ☐
A tourist is inquiring someone about information of attractions ☐
They are in the airport ☐

2. Listen and complete the blanks

| Dynasty | one of | recommend | hill | wait | worth |
| traveling | proud | relics | attractions | Pagoda | enjoy |

A: Excuse me, sir. I'm here for _____ and I think I'm kind of lost here.

B: Okay. Where do you want to go?

A: Well ... Do you have any good ideas about seeing _____? Cause it's my first time be here.

B: Oh, I'm very _____ to tell you something about Xi'an. It's _____ China's oldest cities. There are many _____, and most of them are very old. But I really like to _____ Terracotta army. It is a super large collection of life-size terra cotta sculptures in battle formations, reproducing the mega imperial guard troops of Emperor Qin Shi Huang, the first emperor of the first unified _____ of Imperial China. It is not far from here. You can go there by bus after reaching Wulukou (五路口) by Xi'an Metro Line 1.

A: What a long history! It is _____ visiting. Is there anything else?

B: Sure, for example, Big Wild Goose (大雁塔)_____, Ci'en Temple (慈恩寺), Small Wild Goose Pagoda (小雁塔), Lishan (骊山)_____, Xi'an City Wall (西安城墙) and Huashan (华山).

A: Sounds very interesting. I'd _____ them. I can't _____ to go there.

B: Wish you have a good time.

3. Words and expressions

emperor	['empərə(r)]	n.	皇帝	pagoda	[pə'gəʊdə]	n.	塔，宝塔
dynasty	['dɪnəstɪ]	n.	朝代	relic	['relɪk]	n.	文物，遗迹
unify	['juːnɪfaɪ]	v.	统一	I can't wait to ...			迫不及待

Lesson 4　Tourist Attractions 旅游景点

Name:_____　Class:_____　Date:_____

4. Read after the recording

Conversations

1. Vocabulary preview

Crane

It's a very beautiful bird with small head and long neck, it has a long and straight mouth and a pair of long and thin legs.

Lee was taking a picture of a group of cranes in the zoo.

East

Sun rises from that direction.

Lucy is planning to spend her summer vacation in the east part of Japan.

Starve

Very hungry.

"When is the dinner ready? Daddy, I'm starving."Cindy asked.

2. Dialogues

Scene A: A passenger is asking about tourist attractions (1)

T-Tina (Ticket staff); H-Helen (Service agent)

H: Hi, would you please recommend some places for sightseeing nearby metro stations?

T: Yes, There are a lot of tourist attractions and resources in Wuhan Metro Line 4, for example, Chu River and Han Street (楚河汉街), which is developed as a project of the Phase 1 of Wuhan Central Cultural Zone. You can take the Metro Line 4 to the destination.

Tina　　　　Helen

H: That sounds great, do you know other attractions? How to get to Wuhan University?

T: Yes, I know it's very close actually. When you arrive at Jiedaokou (街道口)station by Wuhan Metro Line 2, you go straight down this road, and then turn left at the next intersection, you can see Wuhan University in your sight. This is one of the most beautiful universities in the world.

H: I am so excited to catch sight of these attractions. That's very kind of you.Thank you very much.

T: With pleasure.

Scene B:A passenger is asking about tourist attractions (2)

A-Allen (Metro intern);C-Cherie (passenger)

C: Would you please recommend some tourist attractions nearby metro stations?

A:Yellow Crane Tower. It is one of the three nationwide famous ancient towers in China, which is located on Snake Hill in Wuchang district (武昌区), it was first built during the Three Kingdoms Period as a military observation

Allen　　　　Cherie

Lesson 4 Tourist Attractions 旅游景点

Name:_____ Class:_____ Date:_____

tower. You can get the top floor to enjoy a distant view of the beautiful Wuhan city and feel the powerful of Yangtze River.

C: Em ... that's the charm of Yellow Crane Tower and the best spot for pictures. Any other suggestions?

A: East Lake. It is the biggest city lake in China which is six times as large as the West Lake in Hangzhou.

C: Unbelievable!

A: It's a big place full of fun and great views. You can go to East Lake Ocean Park to feed animals, enjoy the "city tree of Wuhan city" and "Moon Enjoying Pavilion". You can also rent a bike riding along the lakefront to enjoy the wonderful view.

C: What's that "city tree" ?

A: Metasequoia. It is a kind of ancient tree which lived in the Quaternary Glacial Period, and it has a history of 130 million years, so we call it "the living fossil". Metasequoia can just symbolize the spirit of Wuhan people's honesty, frankness and firmness.

C: What is "Moon Enjoying Pavilion"?

A: It is an ideal place for you to enjoy the moon. It got the name from the Chinese poem *The nearer a pavilion lies to water, the earlier someone standing on the pavilion can enjoy the moon*. We can have a stroll along the East Lake to enjoy the magnificent view of the pavilion.

C: Wow ... that's great. East lake sounds like a good place to go.

A: Yes. And also delicious foods nearby.

C: Which line should we take?

A: Line 8. You can get on at Zhaojiatiao (赵家条) station and get off at terminal station. And get out from Exit C. It only takes 30 minutes to get there.

C: Okay. Thanks.

Scene C: A passenger is requiring a way to a local attraction

H-Helen (Service agent); P-Peter (Passenger)

P: I'm sorry to bother you, madam. I'm here for traveling and now I want to go to Disney, I'm not sure which route is correct. Would you please give me a hand?

H: Shoot.

Helen   Peter

P: I was searching on my phone. It shows if I want to go to Disney, I need to go to Longyang (龙阳) Road first.

H: Yeah, it is. Longyang Road is an interchange station. You are lucky because metro Line 2 could take you there, and we are now at Line 2.

P: Then I need to transfer to Line 16 and get off at Dishui Lake (滴水湖) after 2 stops.

H: Well ... Just give me a second. I think you should get off at Luoshan (罗山) which is also a transfer station and transfer to Line 11, after 2 stops you will arrive at Disney. That would be faster.

P: Okay. It doesn't matter. Thank you anyway. I have my phone and I can also ask someone else on the way to Disney.

Lesson 4 Tourist Attractions 旅游景点

Name:_____ Class:_____ Date:_____

H: Yes, you are right. I think you could make it. Have a good time in Shanghai.

P: Thank you, I will. Bye.

Scene D: A service agent and her friend are talking about the itinerary

H-Helen (Service agent); B-Helen's friend

H: I'm starving ...

B: Yeah. I'm also hungry. We were shopping at Coco Park for whole afternoon. Let's order something to eat.

Helen

H: Sure. Let me scan the code to order our dishes.

(After a few minutes ...)

B: Any plan for tomorrow?

H: Windows of the World?

B: Wow ... Great! That's the place where exactly I want to visit.

H: It's a cultural scenic spot and contains world-famous landscapes.

B: Yeah, that's a good place for taking pictures.

H: Anyway. I think "Windows of the World" is a terrific place. Let's go there tomorrow. I am sure it will astonish us.

B: I just checked on my phone, the map shows that metro Line 1 and Line 2 could take us to get there.

H: That's quite convenient for us cause the hotel we checked-in is near the metro station.

B: Tell me about it …

3. Words and expressions

right away			立刻，马上	interchange station			换乘站（transfer station）
Yellow Crane Tower			黄鹤楼	itinerary	[aɪˈtɪnərərɪ]	n.	旅行计划，行程安排
East lake			东湖	Windows of the World			世界之窗
full of			充满，装满，充斥	scan	[skæn]	v.	扫描
lake	[leɪk]	n.	湖	scan the code			扫码
metasequoia	[ˌmetəsɪˈkwɔɪə]	n.	水杉	world-famous			举世闻名
lakefront	[ˈleɪkfrʌnt]	n.	湖边	landscapes	[ˈlændskeɪps]	n.	风景
stroll	[strəʊl]	v.	散步，闲逛，溜达	take picture			照相
feed	[fiːd]	v.	给……食物吃，喂	terrific	[təˈrɪfɪk]	adj.	极好的，了不起的
correct	[kəˈrekt]	adj.	对的，正确的	astonish	[əˈstɒnɪʃ]	v.	使惊讶

Lesson 4　Tourist Attractions 旅游景点

Name:_____　Class:_____　Date:_____

4. Useful sentences and phrases

How about going to Yellow Crane Tower after breakfast?　早餐后我们去黄鹤楼怎么样呢?

How about ...　怎么样?

We can also rent a bike riding along the lakefront to enjoy the wonderful view.　我们还可以租辆自行车沿着湖边骑，欣赏美丽的风景。

We can also ...　我们还可以……

rent a bike　租辆自行车

enjoy the wonderful view　欣赏美景

East lake sounds like a good place to go.　东湖听起来是个好去处。

It sounds like a good/nice/wonderful/interesting/place to go. ...　听起来是个好去处。

For example: *Taihu Lake* (太湖) *sounds like a wonderful place to go.*

I'm here for travel.　我是来旅游的。

I'm here for ...　我在这里为了……

For example: *I'm here for study. /I'm here for a business trip./I'm here for fun ...*

Have a good time in Shanghai.　祝你在上海玩得愉快。

have a good time in　在……玩得开心。

For example : *Have a good time in the party. /Have a good time in the meal.*

I'm starving.　我饿极了!

5. Grammar

Come out to have your breakfast.

介词to的用法：后面跟名词或动名词，表示结果、目的、对象、原因等。"Come out to have your breakfast."表示目的，即出来是为了吃早餐。

Let's go somewhere else.

else和other用法区别：需要注意它们的所有格。else是个副词，常见的用法是：一是只用来修饰两类词；二是必须放在被修饰词的后面。For example: *Would you like something else to eat?/Ask someone else to help you./What else do you want?* other是个形容词，修饰名词或代词，用在名词或代词之前。For example: *Do you have any other questions?/I have two cats. One is black, the other one is orange.*

Let's go to East Lake then.

then *adj.* 当时的；*n.* 然后；*adv.* 那时，当时。

在本句中then用作副词。

Hilda took her bag and then left home to school.　在此句中then为名词，有"然后"的意思。

We lived in Tokyo then.　我们那个时候住在东京。在此句中then为副词，有"那时"的意思。

That's the place where exactly I wanted to visit.

Exactly *adv.*　正好,刚好；精确地，确切地。

在本句中exactly表示"正好，刚好"的意思。那个地方就是我想去的。*The airplane arrived exactly at 4 p.m.*　飞机下午4点整到达。

Lesson 4　Tourist Attractions 旅游景点

Name:_____　Class:_____　Date:_____

6. Exercises

Task 1: Fill in the blanks with the words from conversations that match the meanings of sentences.

(1) _____ is the first meal in the day.

(2) _____ is a place where you can see all sorts of ocean animals.

(3) _____ basically means "right".

(4) _____ means super famous, everyone in the world knows it.

(5) _____ means "immediately".

Task 2: Fill in blanks with the given words

| full of | lakefront | code | check in | wonderful | world-famous |

(1) I was walking by the _____ with my mother this afternoon.

(2) Please scan the _____ to pass the turnstile.

(3) The metro station is _____ people at the rush hour.

(4) I'm having a _____ time in Ada's birthday party.

Task 3: Translate the following sentences into English

（1）今年寒假我想去上海迪士尼玩。

（2）我来这里（地铁车站）是实习的。

（3）你能告诉我哪条地铁到小雁塔吗？

（4）请您扫二维码出站。

（5）我马上处理这个问题。

（6）哇！这里的景色真的是太美了。

Task sheet

Objectives

Learn how to describe a tourist attraction.
Learn how to introduce seeing attractions correctly.

Oral task of grading details

Specific requirements for the academic quality of College English in higher Vocational Education.

Levels（水平分类）	Qualitative description （质量描述）
Level 1 (general requirements)〔水平一（一般要求）〕	Level 1-1: Can basically understand clear pronunciation, slow speech in daily life and urban rail transit passenger service posts related topics〔能基本听懂发音清晰、语速较慢的日常生活语篇和职场（城市轨道交通客运服务岗位）相关话题〕

Lesson 4 Tourist Attractions 旅游景点

Name:_____ Class:_____ Date:_____

Continued

Levels（水平分类）	Qualitative description（质量描述）
Level 1 (general requirements)〔水平一（一般要求）〕	Level 1-2: Can basically understand the relevant English materials of urban rail transit passenger transport service, understand the main contents and obtain key information; understand the cultural connotation and identify the professional terms of urban rail transit passenger transport service（能基本读懂、看懂城市轨道交通客运服务的相关英语资料，理解主要内容，能获取关键信息；领会文化内涵，能识别城市轨道交通客运服务的专业术语）
	Level 1-3: Can communicate with others on familiar topics in daily life and urban rail transit passenger transport service; the expression is basically accurate and fluent, and can briefly introduce workplace culture and METRO culture（能在日常生活中和城市轨道交通客运服务工作中就比较熟悉的话题与他人进行语言交流，表达基本准确、流畅；能简单介绍职场文化和地铁企业文化）
	Level 1-4: Can briefly express their experiences, opinions and feelings; the sentences are basically correct and expressed clearly（能简要表达自己的经历、观点、感受；语句基本正确，表达清楚，格式恰当）
	Level 1-5: Can meet the basic communication needs on familiar with daily life and urban rail transit passenger transport service（能就日常生活和城市轨道交通客运服务工作中熟悉的话题，满足基本沟通需求）
	Level 1-6: Be able to make a clear learning plan; to obtain learning resources through online and offline channels under the guidance of teachers（能制订明确的学习计划；能在教师引导下通过线上线下多种渠道获取学习资源）

Tasks

Task 1. Describe the pictures
Task 1-1: Disney

💡 **Way out**

Have you ever been there?
Do you like this place? Do you want to go to this place again?

💡 **Language notes**

station　　车站
供列车停靠、乘客购票、候车和乘降并设有相应设施的场所。
at grade station　　地面车站
轨道设在地面上的车站。

Lesson 4　Tourist Attractions 旅游景点

Name:_____　Class:_____　Date:_____

underground station　　地下车站
轨道设在地面下的车站。
专业术语来源：《城市轨道交通工程基本术语标准》（GB/T 50833—2012）
（*Standard for basic terminology of urban rail transit engineering*）

💡 **Outlines**

_____

_____

_____

Task 1-2: The Great Wall

What's this?
Have you ever been there ?
How to introduce this relic to the foreign passengers?

💡 **Notes**

The Great Wall reflects collision and exchanges between agricultural civilizations and nomadic civilizations in ancient China. It provides significant physical evidence of the far-sighted political strategic thinking and mighty military and national defense forces of central empires in ancient China, and is an outstanding example of the superb military architecture, technology and art of ancient China. It embodies unparalleled significance as the national symbol for safeguarding the security of the country and its people.

💡 **Outlines**

_____

_____

_____

Lesson 4   Tourist Attractions 旅游景点

Name:_____   Class:_____   Date:_____

Task 2. Practice your oral English and Role play

Task 2-1: Scene A

Work in pairs. Suppose you were having a trip in Hong Kong and you want to go to Disney.

💡 **Useful expressions/patterns**

Excuse me. I want to go to ... would you please tell me ... / I was wondering if you could tell me ... / I'm not sure the way I'm heading to ... is correct, would you please show me the way?

Sure. It's not far from here.../ Of course. You just need to...

💡 **Outlines**

_____

_____

_____

Task 2-2: Scene B

Work in pairs. Suppose you and your friend were talking where to spend your holiday.

💡 **Useful expressions/patterns**

Any plan for tomorrow? / I was wondering if we could go to...for our holiday.

Sure . / Okay, I'll go with you ... / How about ... /That's the place where exactly I wanted to visit.

💡 **Outlines**

_____

_____

_____

Self-evaluation

Item	Excellent （90~100分）	Good （80~90分）	Average （60~80分）	Pass （60分）	Fail （<60分）
Be able to use phrases to make conversations					
Take part in pair work and role play					
Improve oral communication skills					
Be able to finish the exercises independently					
Review and preview lessons consciously					

Lesson 4　Tourist Attractions 旅游景点

Name:_____　Class:_____　Date:_____

Grade evaluation

Item（项目）	No.（序号）	Criteria（标准）	Score（分数）		
			Self-evaluation（自评）	Peer（学生互评）	Facilitator（指导教师）
General Evaluation（一般评价）（25%）	1	Teamwork, well-assigned jobs（能进行团队合作，能做到分工良好）			
	2	Group discussions for the topic, plan for oral task materials（能进行小组讨论，能准备口语训练任务材料）			
	3	Well-prepared（准备充分）			
	4	Role playing well-performed（角色扮演表现良好）			
	5	Self-confidence（自信）			
Evaluation of Professional Competence（专业能力评价）（25%）	1	Pronunciation, articulation（发音清晰）			
	2	Accuracy, fluency（语言准确、流利）			
	3	Tone of voice, coherent（语调连贯）			
	4	Ability of cross-cultural communication（跨文化交流能力）			
	5	Presentation skills（口语技巧）			
Task-based Evaluation（基于任务的评估）（50%）	1	Describe the pictures（描述图片）			
	2	Practice your oral English and Role play（英语口语练习与角色扮演）			
Final Score（合计）		Self-evaluation Score × 20% + Peer Score × 30% + Facilitator Score × 50%（学生自评分数×20%+学生互评分数×30%+教师点评分数×50%）			
Facilitator's Comments（教师评价）					

Task reflection

FACT: What do you get?

FEELING: How do you feel?

FINDING: What do you find?

FUTURE: What shall you do next?

Lesson 4　Tourist Attractions 旅游景点

Name:_____　Class:_____　Date:_____

Complementary reading

Interesting Subway Stations

Interesting stations dedicated to public transportation can be found all over the world. A subway (or metro) station is considered interesting either by its efficiency or by its beautiful architectural designs and art work. These subways not only offer transport from one point to another but in some cases, also qualify as a tourist attraction. They improve the aesthetic value of the town or city in which they are built.

1. Moscow—Arbatskaya

Arbastskaya is a subway station on the Arbatsko-Pokrovskaya Line of the Moscow Metro. It was designed by Leonid Polyakov, Valentin Pelevin, and Yury Zenkevich. The main tunnel is elliptical, not the usual circular design. It has square pylons with a red marble finishing. The ceiling is decorated with ornamental brackets, floral reliefs and patterns, and beautiful chandeliers to provide lighting.

2. Stockholm—T-Centralen

T-Centralen subway station has rugged finishing which brings out that cave-like look for which Sweden's Stockholm Metro is so pretty. It is the meeting point of the 3 Stockholm Metro lines. T-Centralen is the most used subway station in Stockholm, and was opened in 1957. However, the blue vines and flower motifs decorations were installed in 1975 when the metro station was being renovated.

3. Lisbon—Olaias Station

Olaias station is a subway station on the Red line of the Lisbon Metro in Lisbon, Portugal. It was designed by Tomas Taveira. The underground metro station opened on May 19, 1998. The finishing of the station, especially on the sides, the ceiling, and the floor makes it one of the most attractive public transit stations in the world. The colorful tiles and brightly colored glass are fixed on the sides and floor, giving it an eye-popping finish.

4. Moscow—Prospekt Mira

5. Stockholm—Stadion Station

6. Saint Petersburg—Avtovo

7. Naples—Toledo

8. Kiev—Zoloti Vorota

9. Paris—Arts Et Metiers

Lesson 4　Tourist Attractions 旅游景点

Name:_____　Class:_____　Date:_____

Translation:

(1) public transportation

(2) A subway (or metro) station is considered interesting either by its efficiency or by its beautiful architectural designs and art work.

(3) from one point to another

(4) a tourist attraction

(5) aesthetic value

(6) in some cases

 Relevant knowledge

Chinese-operated light rail completes hajj mission

　　A shuttle train runs on the elevated light rail in Mecca, Saudi Arabia. The Mecca Metro, built and operated by China Railway Construction Corp, proved a huge success during the recently concluded annual hajj pilgrimage that saw nearly 2 million visitors this year.

　　Millions of pilgrims shuttled in cool comfort between holy sites during the recently concluded annual hajj pilgrimage in Saudi Arabia, thanks to a light rail service built and operated by China.

　　The Mecca Metro, also known as the Makkah Metro, which was built and run by State-owned China Railway Construction Corp, ferried pilgrims between holy sites, helping ease road traffic congestion during the annual event and winning widespread praise.

(Source: China Daily | 2023-07-03)

Lesson 4　Tourist Attractions 旅游景点

Name:_____   Class:_____   Date:_____

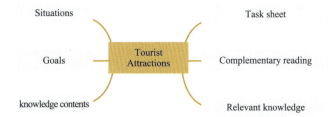

Part Three
Station Services

第3部分　站　务　服　务

Smart stations make travel more convenient.
智慧车站使出行更便捷。

  Nowadays, more stations have become a lot smarter. For example, an intelligent customer service robot offers multi-lingual services, Ticket Vending Machines support voice recognition and passengers can make use of a facial recognition system to board and pay fares without having to remove their face masks. These smart stations provide passengers with greater travel convenience.

  当今，越来越多的地铁车站变得更智能。例如，车站智能客服机器人可提供多语言服务，TVM自动售票机增加了语音功能，过闸机时有智慧人脸识别系统，戴口罩也可刷脸通行。这些智能车站让乘客出行更便捷。

Lesson 5　　Tickets and Fares
第5课　车票与票价
Lesson 6　　Checking Tickets
第6课　检票
Lesson 7　　Station Facilities
第7课　车站设备
Suggested hours: 6-8 class hours

Lesson 5 Tickets and Fares
第5课 车票与票价

城市轨道交通客运服务英语口语（第3版）

📖 Overview

Situations (学习情境)

Goals (学习目标)

Groups (任务分组)

Knowledge contents (知识储备)

    Lead-in (导入)

    Listening (听力)

    Conversations (情境对话)

Task sheet (口语训练单)

    Objectives (训练目标)

    Tasks (训练内容)

    Self-evaluation (自我评价)

    Grade evaluation (成绩评定)

    Task reflection (任务反思)

Complementary reading (拓展阅读)

Relevant knowledge (知识链接)

Summary (总结)

📖 Situations

- A passenger is buying a ticket for an individual
- A passenger is buying tickets for a group
- A passenger is buying tickets for kids

    Q：假设你是一名售票员（Ticket staff），你将如何为乘客提供购票信息？你会如何介绍地铁车票的种类和票价？如何用英语开口表达呢？

Lesson 5　Tickets and Fares 车票与票价

Name:_____　Class:_____　Date:_____

🎓 Goals

- Provide tickets information
- Master terminology
- Improve oral communication skills
- Improve professionalism and service awareness

🎓 Groups

Class（班级）		Group No.（小组编号）	
Leader（组长）		Facilitator（指导老师）	
Members（组员）		Roles（角色分工）	

🎓 Knowledge contents

Lead-in

1. match the words and pictures

MTR　　　　　　　　　Payment　　　　　　　Octopus Card

Picture A

Picture B

Picture C

2. Match the pictures and sentences

Picture A

(1) It is a stored value card that provides customers to take MTR.

Picture B

(2) It is the transfer of an item of value from a person/company to another in exchange for goods or services.

Picture C

(3) Mass Transit Railway is the mass transit system of Hong Kong, China.

Lesson 5   Tickets and Fares 车票与票价

Name:_____   Class:_____   Date:_____

3. Discuss and share

💡 Do you know something about Octopus Card? If you do, tell us about it.

Listening

1. Listen and tick the topic

The charge of MTR ☐
A trip to Hong Kong ☐
Meeting with an old friend ☐

2. Listen and complete the blanks

| fare | bug | that's a point | convenience | Octopus Card |
| admission notice | congratulations | Airport Express | concessionary |

A: I am sorry to _____ you, since I am new here, and I want to know the ticket _____, could you please tell me something about it?

B: Certainly. There are three different fare classes on the MTR (Mass Transit Railway): adult, students and _____.

A: What's the difference?

B: The concessionary rate is only for children (under 12) and the elder (over 65) to use, full-time Hong Kong students between the ages of 12-25 qualify for the concessionary rate using an Octopus Card on all lines except on _____. And there are three kinds of payment: _____, single journey ticket and passenger ticket.

A: Okay ... I got it.

B: Well ... You look so young. Are you a student?

A: Yes, I am. I am here for studying; I've just got the _____ from the University of Hong Kong.

B: Wow ... That's wonderful. _____! Then you can use the concessionary ticket, and I suggest you to buy an Octopus Card, it's rather _____.

A: _____. I'll take your advice. Thanks a lot.

B: You're welcome.

3. Words and expressions

I am sorry to bug you			不好意思打扰一下
concessionary	[kən'seʃənərɪ]	adj.	（对某人）减价的
admission notice			录取通知书
Octopus Card			八达通卡
Airport Express			机场快线
all kinds of (all sorts of)			各式各样
full-time	['fʊl 'taɪm]	adj.	全职的；全日制的

Lesson 5 Tickets and Fares 车票与票价

Name:_____   Class:_____   Date:_____

congratulation	[kənˌgrætjʊˈleɪʃən]	n.	祝贺
take	[teɪk]	v.	接受
That's a point (good idea/good thought).			好主意!

4. Read after the recording

Conversations

1. Vocabulary preview

Ticket office
It is the office where tickets of admission are sold.
There is a long queue outside the ticket office.

Discount
It is a reduction in the usual price of something.
I can give you a small discount.

Token
It is a round piece of plastic used instead of money to take metro or as a form of payment.
You could go to the TVM and use some coins to get a token.

2. Dialogues

Scene A: A passenger is buying a ticket for an individual
T-Tina (Ticket staff); M-Mr. Thomas (passenger)

M: Excuse me, would you please tell me how I can buy a single journey ticket?

T: First of all , I am glad to offer help. You can go to the ticket office to get one.

M: Thank you, Miss. Where is the ticket office?

T: Look, over there. It is on the right of the ticket barrier. Please stand in line to buy a metro card.

Tina   Thomas

M: Oh, there are so many people waiting in line.

T: Maybe you could go to the TVM and use some coins to get a token. Let me show you. This way, please ...

M: OK, thanks.

T: First, exchange note into coins at the ticket booth, cause it only accepts 1 yuan, 5 yuan, 10 yuan note. Then go to TVM, choose the metro line and your destination. Finally, you can get the token.

M: What should I do with the token?

T: You put it in the slot at the turnstile and then push the turnstile to get into the platform.

M: It sounds a bit complicated.

T: That may sound complicated, but it is simple. Or, you could buy a store value ticket.

M: How can I get one?

Lesson 5　Tickets and Fares 车票与票价

Name:_____ 　Class:_____ 　Date:_____

Tina　　　Cherie

T: You should pay a 20 yuan deposit.

M: Thank you so much. You are so helpful.

T: You are welcome. Have a good trip!

Scene B: A passenger is buying tickets for a group

T-Tina (Ticket staff); C-Cherie (Passenger)

C: How many kinds of current tickets in Beijing Metro Line 2?

T: There are five kinds of current tickets in Beijing Metro Line 2: single journey ticket, store value ticket, student ticket, elder ticket and group ticket.

C: How can I get a group ticket, and how about the discount?

T: We only offer a group ticket for 20 persons, you can contact with the station staff, the group more than 20 persons, not exceed 100 persons may have 10 percent discount; more than 100 persons may have 20 percent discount.

C: OK, we have 30 persons. How much do I owe you?

T: 30 persons for a group ticket, altogether 405 yuan.

C: OK, here is 410 yuan.

T: There are 30 tickets and change. Have a good trip!

Scene C: A passenger is buying tickets for kids

T-Tina (Ticket staff); P-Peter (Passenger)

P: Excuse me, should I buy tickets for my kids?

T: How tall are they?

P: My boy is about 1.1 meters and my girl is nearly 1.4 meters tall.

Tina　　　Peter

T: According to our policy, children below 1.2 meters in height ride for free when accompanied by a paying adult. Students who are over 1.2 meters tall can buy student tickets and enjoy 20% discount.

P: OK, thank you.

T: Please take care of your kids.

P: Thank you for reminding me.

T: Good bye!

3. Words and expressions

ticket barrier			检票口	store value ticket		储值票
TVM (Ticket Vending Machine)			自动售票机	student ticket		学生票
deposit	[dɪˈpɒzɪt]	n.	押金	elder ticket		敬老票
single journey ticket			单程票	group ticket		团队票

4. Useful sentences and phrases

Would you please tell me how I can buy a single journey ticket?　　请告诉我怎样买单程车票?

How many kinds of currents tickets in Beijing Metro Line 2?　　目前北京地铁2号线有几种类型的票?

Lesson 5 Tickets and Fares 车票与票价

Name:_____   Class:_____   Date:_____

Please stand in line to buy a metro ticket.　　请排队购买地铁票。
There are 30 tickets and change. Have a good trip!　　这是30张票和找零。祝您旅途愉快！
How much do I owe you?　　我该付你多少钱？
It sounds a bit complicated.　　听起来有点复杂。
According to our policy ...　　根据我们的政策……
Thank you for reminding me.　　谢谢你的提醒。

5. Grammar

stand in line 和 stand in a line
stand in line　　　排队
stand in a line　　站成一排
How much is the fare? / How much is this? / How much does this cost? / What is the price of this?　　价格是多少？
Could you tell me how much it is?　　你能告诉我价格是多少吗？
How much is the fare?　　车费是多少？
What's the room rate?　　房价是多少？

6. Exercises

Task 1: Fill in the blanks with the words from conversations that match the meanings of sentences

(1) _____ is something such as a fence or wall that is put in place to prevent people from moving easily from one area to another.

(2) _____ means happening, being used, or being done at the present time.

(3) A _____ is a person who assist his superior in carrying out an assigned task.

(4) If something is _____, that means it has so many parts or aspects that it is difficult to understand or deal with.

(5) _____, it is a kind of ticket for a single trip.

(6) If you _____ someone, you go somewhere with them.

(7) A _____ is a sum of money which you pay when you start renting something.

Task 2: Fill in blanks with the given words

| complicated | barrier | meters | metro card | discount |
| in line | owe | simple | | |

(1) The ticket office is on the right of the ticket _____.

(2) Excuse me, where can I buy a _____?

(3) I need two tickets, how much do I _____ you?

(4) Excuse me, should I buy a ticket for my child? He is only 1.3 _____.

(5) There are 45 persons in my group, could you give a _____ for me?

(6) Oh, there are so many people waiting _____.

(7) That may sound _____, but it is _____.

Task 3: Translate the following sentences into English

（1）请问我在哪里可以买地铁票？

Lesson 5　Tickets and Fares 车票与票价

Name:_____　Class:_____　Date:_____

（2）请问小孩需要买票吗？

（3）请问地铁5号线有几种票价？

（4）怎么购买敬老票？

（5）根据规定，由成人陪同的身高不足1.2米的儿童免票。

（6）请照顾好您的小孩。

 Task sheet

 Objectives

Learn how to describe a metro ticket (color、size、function).
Learn how to describe the advantages of metro cards.

Oral task of grading details

Specific requirements for the academic quality of College English in higher Vocational Education.

Levels（水平分类）	Qualitative description（质量描述）
Level 1 (general requirements)［水平一（一般要求）］	Level 1-1: Can basically understand clear pronunciation, slow speech in daily life and urban rail transit passenger service posts related topics［能基本听懂发音清晰、语速较慢的日常生活语篇和职场（城市轨道交通客运服务岗位）相关话题］
	Level 1-2: Can basically understand the relevant English materials of urban rail transit passenger transport service, understand the main contents and obtain key information; understand the cultural connotation and identify the professional terms of urban rail transit passenger transport service（能基本读懂、看懂城市轨道交通客运服务的相关英语资料，理解主要内容，能获取关键信息；领会文化内涵，能识别城市轨道交通客运服务的专业术语）
	Level 1-3: Can communicate with others on familiar topics in daily life and urban rail transit passenger transport service; the expression is basically accurate and fluent, and can briefly introduce workplace culture and METRO culture（能在日常生活中和城市轨道交通客运服务工作中就比较熟悉的话题与他人进行语言交流，表达基本准确、流畅；能简单介绍职场文化和地铁企业文化）
	Level 1-4: Can briefly express their experiences, opinions and feelings; the sentences are basically correct and expressed clearly（能简要表达自己的经历、观点、感受；语句基本正确，表达清楚，格式恰当）
	Level 1-5: Can meet the basic communication needs on familiar with daily life and urban rail transit passenger transport service（能就日常生活和城市轨道交通客运服务工作中熟悉的话题，满足基本沟通需求）
	Level 1-6: Be able to make a clear learning plan; to obtain learning resources through online and offline channels under the guidance of teachers（能制订明确的学习计划；能在教师引导下通过线上线下多种渠道获取学习资源）

Lesson 5　Tickets and Fares 车票与票价

Name:_____　Class:_____　Date:_____

Task 1. Describe the pictures
Task 1-1: Ticket

💡 **Way out**

What kind of ticket is it?

What's the discount on this ticket?

How much is the deposit?

💡 **Language notes**

ticket system　　票制

城市轨道交通乘车收费制度。即车票分类、制作、发售、使用规则及计费方法、票价率等规定的总称。

single journey ticket　　单程票

仅在一次进出站的乘行中有效的车票。

专业术语来源：《城市轨道交通工程基本术语标准》（GB/T 50833—2012）（*Standard for basic terminology of urban rail transit engineering*）

💡 **Outlines**

_____

_____

_____

Task 1-2 Ticket

Lesson 5   Tickets and Fares 车票与票价

Name:_____   Class:_____   Date:_____

💡 Way out

What kinds of tickets are these?
What kinds of persons are suitable for these tickets?

💡 Outlines

Task 2. Practice your oral English and Role play

Task 2-1: Scene A

Work in pairs. Suppose you were buying tickets at the Ticket Office, someone jumps the queue.

💡 Useful expressions/patterns

Greet. /Ask someone to stand in line...
Explain. / Give reasons for that...

💡 Outlines

Task 2-2: Scene B

Work in pairs. Use Alipay to buy a metro ticket at the Ticket Office.

💡 Useful expressions/patterns

Greet. / Want to buy a metro ticket. / Use Alipay to get one. / Express thanks.
Greet. / Show the types of tickets and give the quotation. / Express good wishes.

💡 Outlines

Task 2-3: Scene C

Work in pairs. Ask for a discount of the ticket.

💡 Useful expressions/patterns

Greet. / Want to buy a metro ticket. / Ask for a discount of the ticket. / Express thanks.
Greet. / Introduce the policy of discounts. / Help the passenger to get one.

Lesson 5　Tickets and Fares 车票与票价

Name:_____　Class:_____　Date:_____

💡 **Outlines**

_____

_____

_____

Self-evaluation

Item	Excellent (90~100分)	Good (80~90分)	Average (60~80分)	Pass (60分)	Fail (<60分)
Be able to use phrases to make conversations					
Take part in pair work and role play					
Improve oral communication skills					
Be able to finish the exercises independently					
Review and preview lessons consciously					

Grade evaluation

Item（项目）	No.（序号）	Criteria（标准）	Score（分数）		
			Self-evaluation（自评）	Peer（学生互评）	Facilitator（指导教师）
General Evaluation（一般评价）（25%）	1	Teamwork, well-assigned jobs（能进行团队合作，能做到分工良好）			
	2	Group discussions for the topic, plan for oral task materials（能进行小组讨论，能准备口语训练任务材料）			
	3	Well-prepared（准备充分）			
	4	Role playing well-performed（角色扮演表现良好）			
	5	Self-confidence（自信）			

Lesson 5　Tickets and Fares 车票与票价

Name:_____　Class:_____　Date:_____

Continued

Item（项目）	No.（序号）	Criteria（标准）	Score（分数）		
			Self-evaluation（自评）	Peer（学生互评）	Facilitator（指导教师）
Evaluation of Professional Competence（专业能力评价）（25%）	1	Pronunciation, articulation（发音清晰）			
	2	Accuracy, fluency（语言准确、流利）			
	3	Tone of voice, coherent（语调连贯）			
	4	Ability of cross-cultural communication（跨文化交流能力）			
	5	Presentation skills（口语技巧）			
Task-based Evaluation（基于任务的评估）（50%）	1	Describe the pictures（描述图片）			
	2	Practice your oral English and Role play（英语口语练习与角色扮演）			
Final Score（合计）		Self-evaluation Score × 20% + Peer Score × 30% + Facilitator Score × 50%（学生自评分数×20%+学生互评分数×30%+教师点评分数×50%）			
Facilitator's Comments（教师评价）					

Task reflection

FACT: What do you get?

FEELING: How do you feel?

FINDING: What do you find?

FUTURE: What shall you do next?

Complementary reading

How to use the metro card / ticket usage

The Online Ticketing Service enables you to pre-order the following Airport Express Tickets and Tourist Tickets, which are available in the form of QR Codes or Tickets. Passengers who choose the QR Code can pass through the gate directly with the code. Passengers opting to redeem the tickets at the station need to go to the Customer Service Centre before entering the gate.

Choose either the QR Code or a Ticket for:
- Airport Express Single Journey Ticket

Lesson 5 Tickets and Fares 车票与票价

Name:_____ Class:_____ Date:_____

- Airport Express Round Trip Ticket
- Airport Express Travel Pass
- Tourist Day Pass
- Tourist Cross-boundary Travel Pass

Translation:

(1) Passengers who choose the QR Code can pass through the gate directly with the code.

(2) Airport Express Single Journey Ticket

(3) Airport Express Round Trip Ticket

(4) Airport Express Travel Pass

(5) The Online Ticketing Service enables you to pre-order the following tickets ...

Relevant knowledge

Different ways to get in and out of the metro station

1. QR code手机扫码进出站

Step 1: Download and install an Alipay application software 支 in your phone.

Step 2: Open the Alipay and then find .

Step 3: Click and choose "Metro" and the code is under below.

Step 4: Face the code to the scan part of gate machine then you will hear a "beep" sound when you go through it.

2. Face recognition人脸识别

When passengers pass the screen, the screen relies on face recognition technology to complete face recognition, and opens the gate to enter the station directly.

乘客经过人脸识别闸机屏幕时，几乎无需停留，屏幕依托人脸识别技术可完成人脸识别，开启闸机，乘客直接进站。

Lesson 5　Tickets and Fares 车票与票价

Name:_____　Class:_____　Date:_____

Summary

- Situations
- Goals
- Knowledge contents
- Tickets and Fares
- Task sheet
- Complementary reading
- Relevant knowledge

Lesson 6 Checking Tickets
第6课 检 票

Overview

Situations (学习情境)

Goals (学习目标)

Groups (任务分组)

Knowledge contents (知识储备)

    Lead-in (导入)

    Listening (听力)

    Conversations (情境对话)

Task sheet (口语训练单)

    Objectives (训练目标)

    Tasks (训练内容)

    Self-evaluation (自我评价)

    Grade evaluation (成绩评定)

    Task reflection (任务反思)

Complementary reading (拓展阅读)

Relevant knowledge (知识链接)

Summary (总结)

Situations

- The security check
- The situation of over-travel
- The situation of time-out

    Q：假如你是一名售票员（Ticket staff），你会如何处理乘客出站时遇到的问题？如何用英语开口表达呢？

Lesson 6　Checking Tickets 检票

Name:_____　　Class:_____　　Date:_____

 Goals

- Provide services for passengers
- Master terminology
- Improve oral communication skills
- Improve professionalism and service awareness

 Groups

Class（班级）		Group No.（小组编号）	
Leader（组长）		Facilitator（指导老师）	
Members（组员）		Roles（角色分工）	

Knowledge contents

Lead-in

1. Match the words and pictures

X-ray baggage scanner　　　　CCTV　　　　Ticket checking

　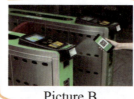　

Picture A　　　　　　Picture B　　　　　　Picture C

2. Match the pictures and sentences

Picture A

　　　　(1) The equipment is for carry-on baggage or luggage, goods, etc. Assistance to detect drug and explosive powder is helpful for security checks.

Picture B

　　　　(2) It is an abbreviation for *closed-circuit television*.

Picture C

　　　　(3) After security check, it is the procedure of carrying a ticket through the barrier into the metro station.

Lesson 6   Checking Tickets 检票

Name:_____   Class:_____   Date:_____

3. Discuss and share

💡 Do you know something about ticket checking? If you do, tell us about it.

Listening

1. Listen and tick the topic

Metro security check ☐
Ticket checking ☐
Security equipment ☐

2. Listen and complete the blanks

| belongings | security | strictly | World Expo |
| X-ray | ticket checking | expanded | CCTV |

In China, metro station _____ checks first started in Beijing ahead of the 2008 Olympic Games, and have remained ever since. In the run-up to the 2010 _____ in Shanghai, the city followed the capital's example. After the World Expo, the metro security checks were _____ to each and every station in the city.

The commonly security equipment consists of _____ baggage scanner, _____, hand-held metal detector and walk-through metal detector.

Personal's belongings, such as bags and _____ should be checked by the X-ray Inspection Instrument in the certain stations. In addition, the following objects are _____ prohibited in the metro station: inflammable, explosive, toxic, corrosive, radioactive, infectious disease pathogen, etc., as well as firearms, ammunition, controlled knives and other articles that may endanger public safety. Regulations for operation management of urban rail transit

Now, due to the impact of COVID-19, the metro station has taken disinfection measures and temperature inspection.

Therefore, passengers should not only accept the temperature test, but also check their _____ before checking in. More and more passengers began to complain that the checks are too time-consuming and the X-ray scanners are too dirty for their treasured handbags.

Although the high security level will waste the time of entering the station and increase the complexity of _____, public health and safety is always the first priority.

3. Words and expressions

in addition	此外	CCTV	n. 闭路电视
consists of ...	由……组成	ticket checking	检票
hand-held metal detector	手持金属探测器	walk-through metal detector	金属探测器（门）
X-ray baggage scanner	X射线行李扫描仪	temperature inspection	体温检测

Lesson 6   Checking Tickets 检票

Name:_____   Class:_____   Date:_____

4. Read after the recording

Conversations

1. Vocabulary preview

Charge
Demand payment.
Will I get charged for this service?

Fare
A fare is the money that you pay for a trip that you make, for example, in a bus or metro.
He could afford the fare.

Check
If you check something such as a piece of information or a document, you make sure that it is correct or satisfactory.
Check your ticket before you get into the station.

2. Dialogues

Scene A: The security check

A-Allen (Metro intern); M-Mr. Thomas (Passenger); C-Cherie (Passenger)

A: Good morning sir, you need to go through the security gate before entering the station.

M: Oh, there aren't any dangerous goods in my bag.

A: Sorry sir, according to our regulations, a passenger who refuses a security check may face criminal detention.

M: Oh, my god! I just want to recharge my card, rather than taking metro.

A: In that case, you can go to the Passenger Service Center.

Allen   Thomas   Cherie

M: Passenger Service Center? Where is it?

A: It's easy to find. You just walk down this hall, the Passenger Service Center is at the end of the hall, and you can see a big sign of it.

M: Okay. I see. Thanks!

A: You're welcome!

(A lady is hurrying to the security gate.)

A: Good morning , Miss. Would you like to go through the security check?

C: It's urgent. I want to go straight.

A: I am afraid you can't. Security check is to ensure the safety of every passenger.

C: But I am in a hurry. I don't have enough time for the security check, and there is nothing dangerous in my handbag.

A: It will take you one or two minutes. Just put your handbag on the belt through the X-ray.

C: OK. By the way, how can I pass through the automatic checking system?

Lesson 6　Checking Tickets 检票

Name:_____　Class:_____　Date:_____

A: Well, just put your ticket in the magnetic area and the gate will open automatically.

C: Thank you so much.

Scene B: The situation of over-travel

T-Tina (Ticket staff); P-Peter (Passenger)

P: Excuse me, could you please tell me why I cannot get out of the station?

T: Please show me your ticket then.

P: Okay. Here you are.

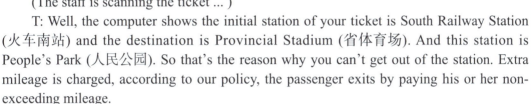

Tina　　　　　Peter

(The staff is scanning the ticket ...)

T: Well, the computer shows the initial station of your ticket is South Railway Station (火车南站) and the destination is Provincial Stadium (省体育场). And this station is People's Park (人民公园). So that's the reason why you can't get out of the station. Extra mileage is charged, according to our policy, the passenger exits by paying his or her non-exceeding mileage.

P: Oh ...What should I do then?

T: Since your ticket balance is insufficient, you simply need to pay the actual exceeding mileage.

P: How much should I recharge?

T: Your ticket fare is 2 Yuan. But the charge from South Railway Station to People's Park will be 4 Yuan. An additional charge is 2 Yuan.

P: Okay. Here is 2 Yuan.

(After a while ...)

T: Alright. It has already done. You can get out of the station now.

P: Thanks for your time!

T: You're welcome! Have a nice day!

Scene C: The situation of time-out

T-Tina (Ticket staff); P-Peter (Passenger); A-Allen (Metro intern)

P: Excuse me, I want to know what happened to my card. Because I can't get out of the station.

Tina　　　　Peter　　　　Allen

A: You can go and ask the Passenger Service Center.

(The passenger now is in the Passenger Service Center.)

P: Excuse me, I can't get out of the station with my metro card. Is there anything wrong with it?

T: Could you please show me your metro card?

P: Certainly. Here you are.

(After seconds ...)

T: Well ... How long did you stay in the metro?

P: More than 4 hours I guess. I was waiting for my friend at the train station.

T: Oh. No wonder! Your ticket had exceeded the system time. According to our ticket policy, rail transit passengers are allowed to stay in the non-free areas for no more than 4 hours for each trip. In case of extra time, 3 yuan (lowest ticket price) shall be additionally

Lesson 6 Checking Tickets 检票

Name:_____ Class:_____ Date:_____

charged before leaving.

P: I got it. Deduct it.

T: Time-out needs to deduct the minimum fare ... Done. Here is your card.

P: Thanks.

T: Good day!

3. Words and expressions

initial	[ɪˈnɪʃəl]	adj.	最初的，初始的	hall	[hɔːl]	n.	大厅，礼堂
destination	[ˌdestɪˈneɪʃən]	n.	目的地，终点	sign	[saɪn]	n.	符号，标志，记号
charge	[tʃɑːdʒ]	n. v.	费用 收费，充，冲锋，指控	pass through/go through			通过
certainly	[ˈsɜːtənlɪ]	adv.	当然，一定	magnetic area			磁力区
policy	[ˈpɒləsɪ]	n.	政策，方针	exceeding mileage			超出里程
deduct	[dɪˈdʌkt]	vt.	扣除				

4. Useful sentences and phrases

What happened to ...? 发生了什么事?
insufficient 余额不足
Passenger Service Center 乘客服务中心
in that case ... 这样的话……
at the end of ... 在……的尽头
no wonder 难怪
According to our policy ... 根据政策规定……

5. Grammar

Would you like to ...?
Could you please ...?
表示请求对方，征询意见的句型。翻译成：你想要/你是否愿意……?
类似的句型还有：
Would you mind ...?
I wonder if you could ...?

6. Exercises

Task 1: Fill in the blanks with the words from conversations that match the meanings of sentences.

(1) If something metal is _____, it acts like a magnet.

(2) _____ is a place where trains can take off.

(3) _____ is a huge place where can hold all sorts of Athletic Races or concerts.

(4) _____ means not enough.

(5) _____ is a place where can offer some services for the passengers.

(6) You use _____ to describe something that happens at the beginning of a process.

(7) If you _____ somewhere, you go there as quickly as you can.

(8) _____ refers to all the measures that are taken to protect a place, or to ensure

Lesson 6　Checking Tickets　检票

Name:_____　Class:_____　Date:_____

that only people with permission enter it or leave it.

Task 2: Fill in blanks with the given words

```
security check     show           destination     stadium
hurry              insufficient   check
```

(1) The Badminton Hall is at the left side of the _____.

(2) Take your time—there's no _____.

(3) Sorry. Your card is balance _____.

(4) Alright! _____ me what you've got.

(5) Where is your _____?

(6) _____ your ticket before you get into the station.

(7) According to our policy, a passenger who refuses a _____ may face criminal detention.

Task 3: Translate the following sentences into English

（1）你可以告诉我为什么我出不了站吗？

（2）我都不知道我的卡余额不足。

（3）这样的话，您可以去客服中心。

（4）根据我们的票务政策，超时需要扣除最低票价。

（5）请你告诉我哪里可以充值。

Task sheet

Objectives

Learn how to assist passengers with tickets checking.
Learn how to describe prohibited items.

Oral task of grading details

Specific requirements for the academic quality of College English in higher Vocational Education.

Levels （水平分类）	Qualitative description （质量描述）
Level 1 (general requirements) ［水平一（一般要求）］	Level 1-1: Can basically understand clear pronunciation, slow speech in daily life and urban rail transit passenger service posts related topics［能基本听懂发音清晰、语速较慢的日常生活语篇和职场（城市轨道交通客运服务岗位）相关话题］

Lesson 6　Checking Tickets 检票

Name:_____　Class:_____　Date:_____

Continued

Levels（水平分类）	Qualitative description（质量描述）
Level 1 (general requirements) ［水平一（一般要求）］	Level 1-2: Can basically understand the relevant English materials of urban rail transit passenger transport service, understand the main contents and obtain key information; understand the cultural connotation and identify the professional terms of urban rail transit passenger transport service（能基本读懂、看懂城市轨道交通客运服务的相关英语资料，理解主要内容，能获取关键信息；领会文化内涵，能识别城市轨道交通客运服务的专业术语）
	Level 1-3: Can communicate with others on familiar topics in daily life and urban rail transit passenger transport service; the expression is basically accurate and fluent, and can briefly introduce workplace culture and METRO culture（能在日常生活中和城市轨道交通客运服务工作中就比较熟悉的话题与他人进行语言交流，表达基本准确、流畅；能简单介绍职场文化和地铁企业文化）
	Level 1-4: Can briefly express their experiences, opinions and feelings; the sentences are basically correct and expressed clearly（能简要表达自己的经历、观点、感受；语句基本正确，表达清楚，格式恰当）
	Level 1-5: Can meet the basic communication needs on familiar with daily life and urban rail transit passenger transport service（能就日常生活和城市轨道交通客运服务工作中熟悉的话题，满足基本沟通需求）
	Level 1-6: Be able to make a clear learning plan; to obtain learning resources through online and offline channels under the guidance of teachers（能制订明确的学习计划；能在教师引导下通过线上线下多种渠道获取学习资源）

Tasks

Task 1. Describe the pictures
Task 1-1: Prohibited items

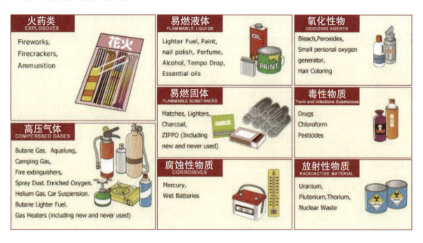

💡 **Way out**

What are the prohibited items?
What are the radioactive materials?
What are the corrosives?
What are the flammable items?

Lesson 6　Checking Tickets 检票

Name:_____　Class:_____　Date:_____

💡 Language notes

safety sign　　安全标志

通过颜色与几何形状的组合表达通用的安全信息，并且通过附加图形符号表达特定安全信息的标志。

service agent　　站务员

在车站从事客运服务工作的服务人员。

ticket staff　　票务员

在城市轨道交通系统内从事票务工作的服务人员。

专业术语来源：《城市轨道交通工程基本术语标准》（GB/T 50833—2012）
（*Standard for basic terminology of urban rail transit engineering*）。

💡 Outlines

_____

_____

_____

Task 1-2: Prohibited & restricted items

💡 Notes

禁止携带下列物品
THE FOLLOWING OBJECTS IS NOT PERMITTED

枪支　　弹药　　警械　　管制刀具　　放射物品
FIREARMS　AMMUNITION　POLICEWEAPONS　CONTROLED KNIFE　RADIOACTIVE

易燃易爆　　腐蚀品　　毒害品　　氧化剂　　强磁物品
FLAMMABLE EXPLOSIVES　CORROSIVES　POISONS　OXIDISING　MAGNETIZED

💡 Outlines

_____

_____

_____

_____

Lesson 6   Checking Tickets 检票
Name:_____   Class:_____   Date:_____

Task 2. Practice your oral English and Role play

Task 2-1: Scene A

Work in pairs. Suppose you were a ticket staff, explain to passengers how to go through the security check.

💡 **Useful expressions/patterns**

Greet. /Ask someone to go through the security check...
Explain. / Give reasons for that...

💡 **Outlines**

_____
_____
_____

Task 2-2: Scene B

Work in pairs. Suppose you were a ticket staff, try to explain to passengers how to check the ticket and get through the gate.

💡 **Useful expressions/patterns**

Greet. / Ask for help.
Greet./Show the way to check the ticket. /Express good wishes.

💡 **Outlines**

_____
_____
_____
_____

Self-evaluation

Item	Excellent (90~100分)	Good (80~90分)	Average (60~80分)	Pass (60分)	Fail (<60分)
Be able to use phrases to make conversations					
Take part in pair work and role play					
Improve oral communication skills					
Be able to finish the exercises independently					

Lesson 6　Checking Tickets 检票

Name:_____　Class:_____　Date:_____

Item	Excellent (90~100分)	Good (80~90分)	Average (60~80分)	Pass (60分)	Fail (<60分)
Review and preview lessons consciously					

Grade evaluation

Item （项目）	No. （序号）	Criteria （标准）	Score（分数）		
			Self-evaluation （自评）	Peer （学生互评）	Facilitator （指导教师）
General Evaluation （一般评价）（25%）	1	Teamwork, well-assigned jobs（能进行团队合作，能做到分工良好）			
	2	Group discussions for the topic, plan for oral task materials（能进行小组讨论，能准备口语训练任务材料）			
	3	Well-prepared（准备充分）			
	4	Role playing well-performed（角色扮演表现良好）			
	5	Self-confidence（自信）			
Evaluation of Professional Competence （专业能力评价）（25%）	1	Pronunciation, articulation（发音清晰）			
	2	Accuracy, fluency（语言准确、流利）			
	3	Tone of voice, coherent（语调连贯）			
	4	Ability of cross-cultural communication（跨文化交流能力）			
	5	Presentation skills（口语技巧）			
Task-based Evaluation （基于任务的评估）（50%）	1	Describe the pictures（描述图片）			
	2	Practice your oral English and Role play（英语口语练习与角色扮演）			
Final Score （合计）		Self-evaluation Score × 20% + Peer Score × 30% + Facilitator Score × 50%（学生自评分数×20%+学生互评分数×30%+教师点评分数×50%）			
Facilitator's Comments （教师评价）					

Task reflection

FACT: What do you get?

Lesson 6　Checking Tickets 检票

Name:_____　Class:_____　Date:_____

FEELING: How do you feel?

FINDING: What do you find?

FUTURE: What shall you do next?

Complementary reading

Metro stations step up anti-virus measures

In 2020, Shanghai Metro stepped up disinfection measures and temperature checks following reports of several local COVID-19 cases.

Seven metro stations at the city's railway stations and Hongqiao (虹桥) International Airport station were being disinfected twice a day instead of once previously, while the Pudong (浦东) International Airport station on Line 2 and the Maglev were disinfected at least four times a day.

Shops inside the stations were disinfected at least as frequently as other parts of the stations.

Liquid soap was again available in all station restrooms, and disposable metro tickets in circulation were disinfected at least once a week, according to the company.

All metro stations were now equipped with thermal imaging devices at security check points to check passengers' temperatures.

Translation:

(1) Shanghai Metro stepped up disinfection measures and temperature checks.

(2) Seven metro stations were being disinfected twice a day instead of once previously.

(3) Liquid soap was again available in all station restrooms.

(4) All metro stations were now equipped with thermal imaging devices at security check points to check passengers' temperature.

Lesson 6 Checking Tickets 检票

Name:_____ Class:_____ Date:_____

📚 Relevant knowledge

Precautions should be taken when taking the metro during the epidemic

1. When taking the metro, wear a mask, do not touch your eyes, mouth and nose, and wash your hands frequently.

搭乘地铁时全程佩戴口罩，不摸眼口鼻，勤洗手。

2. When entering the station please cooperate with the staff to take a temperature inspection.

进站乘车时请配合工作人员进行测温和安检。

3. Pay by Alipay when taking metro to minimize the use of token.

使用支付宝扫码乘车，尽量减少使用筹码式单程票。

4. Keep distance from others on the train.

乘车时保持距离。

5. No conversation、no contact on the train.

大家在搭乘地铁时，不交谈、不接触。

📚 Summary

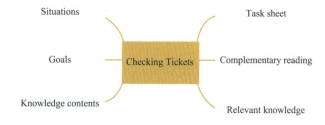

83

Lesson 7 Station Facilities
第7课 车站设备

城市轨道交通客运服务英语口语（第3版）

Overview

Situations (学习情境)

Goals (学习目标)

Groups (任务分组)

Knowledge contents (知识储备)

    Lead-in (导入)

    Listening (听力)

    Conversations (情境对话)

Task sheet (口语训练单)

    Objectives (训练目标)

    Tasks (训练内容)

    Self-evaluation (自我评价)

    Grade evaluation (成绩评定)

    Task reflection (任务反思)

Complementary reading (拓展阅读)

Relevant knowledge (知识链接)

Summary (总结)

Situations

- How to use TVM
- How to use AGM
- Barrier free service facilities
- Other facilities: seats and priority seats

    Q：作为地铁公司员工或地铁公司实习生（Metro staff/ Metro intern）角色，如何引导乘

Lesson 7　Station Facilities 车站设备

Name:_____　Class:_____　Date:_____

客正确使用自动售票机（Ticket Vending Machine）和自动检票机（Automatic Gate Machine）？

📖 Goals

- Introduce station facilities
- Master terminology
- Improve oral communication skills
- Improve professionalism and service awareness

📖 Groups

Class（班级）		Group No.（小组编号）	
Leader（组长）		Facilitator（指导老师）	
Members（组员）		Roles（角色分工）	

📖 Knowledge contents

Lead-in

1. Match the words and pictures

Air condition　　　　　　Seat　　　　　　Handrail

Picture A

Picture B

Picture C

2. Match the pictures and sentences

Picture A

(1) It is a narrow bar that you can hold onto for support.

Picture B

(2) It is an object that you can sit on, for example, a chair.

Picture C

(3) It is for controlling humidity and temperature.

Lesson 7   Station Facilities 车站设备

Name:_____   Class:_____   Date:_____

3. Discuss and share

💡 Do you know something about station facilities? If you do, tell us about it.

Listening

1. Listen and tick the topic

Meeting with an old friend
A trip to Hong Kong
Essential facilities for metro cars

2. Listen and complete the blanks

| facilities | access | calm | handrails | slippery | replaced |
| technician | gray | | air condition | | flooring |

When we enter the car, can you find what _____ are essential for passengers? There are carbon steel seats, _____, passenger information sign, curved ceiling, the _____, the floor heater, and the floor and so on. Look at the interior floor, it has three layers, first the insulation, then the underfloor boards, and later the _____. The surface of the floor is not as _____ as it looks. The floor is made of rubber and clay material which is durable and skid resistance. What's more, the floor is a modular structure which could be _____ quickly and easily.

Here is the heating and air condition. It has the same module as the floor. _____ installs two units on the roof where they are easy to _____ if one breaks down. Workers can just replace them and send a bad one to the manufacturer. The ventilation system is key to passenger's comfort. It circulates clean air throughout the car even in the dark or rainy days. Finally, let's look at the seats. Why are the seats all _____? The gray seats replace the red and yellow seats of the past cars. Because the red color was associated with violence and the new ones can _____ mood. They are very tough and durable ...

3. Words and expressions

essential	[ɪˈsenʃəl]	adj.	基本的；必要的	skid resistance		防滑
carbon steel seats			碳钢座椅	what's more		更重要的是；另外
passenger information sign			信息显示屏	break down		发生故障
curved ceiling			曲形天花板	ventilation system		通风系统
interior	[ɪnˈtɪərɪə(r)]	adj.	内部的，里面的	is associated with ...		和……联系在一起；与……有关
is made of ...			由……所组成	They are very tough and durable.		它们很坚硬也很耐用

86

Lesson 7 Station Facilities 车站设备

Name:_____   Class:_____   Date:_____

4. Read after the recording

Conversations

1. Vocabulary preview

Priority

If you give priority to something or someone, you treat them as more important than anything or anyone else.

Wheelchairs have priority at this junction.

Complex

It has many different parts, and is therefore often difficult to understand.

It's a complex issue, but he only sees it in black and white.

Wear and tear

It is the damage or change that is caused to something when it is being used normally.

The carpet is designed to stand up to a lot of wear and tear.

2. Dialogues

Scene A: How to use TVM

T-Tina (Ticket staff); P-Peter (Passenger)

P: Excuse me, how can I buy a ticket?

T: I am glad to offer help. You can go to the ticket office or TVM to get one.

P: Thank you, Miss. Could you tell me how to use TVM?

T: Sure. There are four steps to use it.

Tina   Peter

P: What are they?

T: First of all, select the right metro line and station of departure and destination. Secondly, press the Fare Selection Button and then deposit coins into the Coin Entry Slot. Finally, you can take the ticket from the Ticket Exit Slot.

P: Thank you so much. You are so helpful.

T: You are welcome. Have a good trip! One more thing, take care of your ticket. We will collect it when you arrive at your destination.

P: Thank you for reminding me.

Scene B: How to use AGM

A-Allen (Metro intern); P-Peter (Passenger)

P: Excuse me, could you tell me how to use the ticket and how to check it?

A: OK, what kind of ticket do you have?

H: Single journey ticket.

Allen   Peter

A: Come here, this is the AGM. Put the ticket in magnetic area, and then you will hear a beep sound; go through as soon as the door opens.

P: It sounds a little complex.

A: If you need help, call for us at any time. Oh, please be sure to keep the ticket when you arrive at your destination.

87

Lesson 7 Station Facilities 车站设备

Name:_____ Class:_____ Date:_____

P: Thanks a lot. How about mobile phone electronic ticket?

A: Open the QR Code, then put the code in sweep area, the door opens after the ticket passes the verification.

P: Thank you.

A: Have a good trip!

Scene C: Barrier free service facilities

A-Allen (Metro intern); B-A man with a wheelchair

A: May I help you, sir?

B: Where is the suitable escalator for my wheelchair?

A: I am afraid that escalator is not safe for you. You'd better switch to the lift.

B: Can you tell me the way to the lift?

A: I will show you there. When you enter the lift, pay attention to the gap.

B: It is so kind of you.

A: Where are you going?

B: I am going to the Yellow Crane Tower, and take some pictures.

A: You should take this elevator to the 2nd basement (B2), ride the train to Guanggu (光谷), and get off at Xiaoguishan (小龟山) station. One more thing, take the elevator out of the station when you arrive at the station.

B: Thank you.

Scene D: Other facilities: seats and priority seats

A-Allen (Metro intern); H-Helen (Service agent)

A: The seats in the car look very heavy, don't they?

Allen   Helen

H: They are tough because they are made of reinforced glass fiber. This material is not only light in weight, but also can stand the wear and tear.

A: Oh, I see.

H: As far as I am concerned the design comes with new belting support, so they are comfortable to sit.

A: What are priority seats?

H: Priority seats are used by passengers using wheelchairs or other mobility devices, or used by passengers with disabilities.

A: How about the elderly or pregnant women?

H: Priority seats may also be used by them.

A: How will I know what seats are designated for priority seats?

H: Seats with different color are designated priority seats. It is generally located near the entrance of train.

A: Do I have to move for a person with a disability?

H: Of course! If you do not have a disability and are sitting in one of the designated priority seats, you must give up your seat for a passenger with a disability, the elderly or a pregnant woman.

Lesson 7　Station Facilities 车站设备

Name:_____　Class:_____　Date:_____

3. Words and expressions

TVM (Ticket Vending Machine)		自动售票机	complex	['kɒmpleks]	adj.	复杂的；合成的
AGM (Automatic Gate Machine)		自动检票机	reinforced glass fiber			玻璃纤维
departure	[dɪ'pɑːtʃə(r)]	n.　离开，离去	wear and tear			磨损
priority seats		优先座	give up			放弃
pregnant women		怀孕妇女	not only ... but also ...			不仅……而且……
QR Code		二维码	car/train		n.	车厢
magnetic area		磁力区				

4. Useful sentences and phrases

Could you tell me how to use the ticket and how to check it?　你能告诉我怎么使用这张票和检票吗？

Priority seats are used by passengers using wheelchairs or other mobility devices, or used by passengers with disabilities.　使用轮椅或其他移动设备的乘客或残疾乘客可使用优先座位。

As far as I am concerned the design comes with new belting support.　就我而言这个设计来自安全带的支撑。

5. Grammar

escalator/elevator/lift
escalator　自动扶梯
elevator/lift　电梯；升降机

英语连词

First of all, select station of departure and destination. Secondly, press the Fare Selection Button, and then deposit coins into the Coin Entry Slot. Finally, you can take the ticket from the Ticket Exit Slot.

（1）表示承接关系：and，and then，too，in addition，furthermore，first，second，third，finally.

（2）表示时间顺序关系：now，then，before，after，afterwards，earlier，later.

（3）表示空间顺序关系：in front of，behind，beside，beyond，above，below.

6. Exercises

Task 1: Fill in the blanks with the words from conversations that match the meanings of sentences

(1) _____ is the place to which they are going or being sent.

(2) _____ are used by passengers using wheelchairs or other mobility devices, or used by passengers with disabilities.

(3) Something that is _____ has many different parts, and is therefore often difficult to understand.

(4) _____ are buildings, pieces of equipment, or services that are provided for a particular purpose.

89

Lesson 7   Station Facilities 车站设备

Name:_____   Class:_____   Date:_____

(5) _____ is the damage or change that is caused to something when it is being used normally.

Task 2: Fill in blanks with the given words

| reminding | one more thing | go through | wheelchair |
| switch | basement | take care of | located |

(1) You are so nice, thank you for _____ me.
(2) You'd better _____ to the lift.
(3) Please use your ticket to _____ the AGM.
(4) Please be sure to _____ the ticket and we will collect it when you arrive at your destination.
(5) It is generally _____ near the entrance of train.
(6) Does the hotel have _____ access?
(7) You should take this elevator to the 2nd _____ (B2).
(8) There's _____ that we can do.

Task 3: Translate the following sentences into English

（1）你能告诉我怎么使用自动售票机吗？

（2）还有一件事，保管好您的车票。

（3）玻璃纤维的材质非常耐磨。

（4）这个听起来有点复杂。

（5）请给需要帮助的乘客让座。

（6）哪里有适合我轮椅的电梯呢？

Task sheet

Learn how to describe the station facilities.
Learn how to use TVM.

Oral task of grading details

Specific requirements for the academic quality of College English in higher Vocational Education.

Lesson 7　Station Facilities 车站设备

Name:_____　Class:_____　Date:_____

Levels（水平分类）	Qualitative description（质量描述）
Level 1 (general requirements) ［水平一（一般要求）］	Level 1-1: Can basically understand clear pronunciation, slow speech in daily life and urban rail transit passenger service posts related topics［能基本听懂发音清晰、语速较慢的日常生活语篇和职场（城市轨道交通客运服务岗位）相关话题］ Level 1-2: Can basically understand the relevant English materials of urban rail transit passenger transport service, understand the main contents and obtain key information; understand the cultural connotation and identify the professional terms of urban rail transit passenger transport service（能基本读懂、看懂城市轨道交通客运服务的相关英语资料，理解主要内容，能获取关键信息；领会文化内涵，能识别城市轨道交通客运服务的专业术语） Level 1-3: Can communicate with others on familiar topics in daily life and urban rail transit passenger transport service; the expression is basically accurate and fluent, and can briefly introduce workplace culture and METRO culture（能在日常生活中和城市轨道交通客运服务工作中就比较熟悉的话题与他人进行语言交流，表达基本准确、流畅；能简单介绍职场文化和地铁企业文化） Level 1-4: Can briefly express their experiences, opinions and feelings; the sentences are basically correct and expressed clearly（能简要表达自己的经历、观点、感受；语句基本正确，表达清楚，格式恰当） Level 1-5: Can meet the basic communication needs on familiar with daily life and urban rail transit passenger transport service（能就日常生活和城市轨道交通客运服务工作中熟悉的话题，满足基本沟通需求） Level 1-6: Be able to make a clear learning plan; to obtain learning resources through online and offline channels under the guidance of teachers（能制订明确的学习计划；能在教师引导下通过线上线下多种渠道获取学习资源）

Tasks

Task 1. Describe the pictures
Task 1-1: Escalator V.S. Lift

小心碰头
Watch your head

请看护好您的小孩
Attend children

小心夹脚
Beware your shoes

请勿推车上行
Troller is not allowed

请小心站稳
Watch your step

请勿在电梯上嬉戏
Do not play on the escalator

请勿携带宠物
Do not bring pets

Lesson 7　Station Facilities 车站设备

Name:_____　Class:_____　Date:_____

💡 Way out

What does the picture describe? How to use it? What are the problems you should pay attention to?

💡 Language notes

自动扶梯　　escalator

带有循环运行梯级，服务于车站规定楼层的向上或向下倾斜运送乘客的固定电力驱动设备。

电梯　　lift/Elevator

服务于车站规定楼层的固定式升降设备，它具有一个轿厢，运行在至少两列垂直的刚性导轨之间，轿厢尺寸与结构形式便于乘客出入。

专业术语来源：《城市轨道交通工程基本术语标准》（GB/T 50833—2012）
（*Standard for basic terminology of urban rail transit engineering*）

💡 Outlines

_____
_____
_____

Task 1-2: How to use AGM?

1. IC card reader (for *meto* card only)
Touch your *metro* card here. The screen will display your remaining balance.

2. Ticket slot (Magnetic tickets)
Insert your ticket into the slot when you get out of the station.

💡 Notes

TVM开站作业程序：打开TVM维护门—补充单程票—补充硬币—装入纸币钱箱—装入硬币回收钱箱并上锁—注销退出—确认TVM设备正常。

TVM关站作业程序：下班盘点—取出票箱—取出1元专用找零钱箱—取出纸币钱箱—取出硬币回收钱箱—运营统计—注销退出—关闭TVM设备。

摘编自：2022一带一路暨金砖国家技能发展与技术创新大赛"城市轨道交通运营设计与应急处理"赛项作业程序。

💡 Outlines

_____
_____
_____

Lesson 7　Station Facilities 车站设备

Name:_____　Class:_____　Date:_____

Task 2. Practice your oral English and Role play

Task 2-1: Scene A

Work in pairs. Introduce TVM to passengers.

💡 **Useful expressions/patterns**

Greet. / Wonder how to use TVM./ Express thanks.
Explain. / Display steps...

💡 **Outlines**

_____
_____

Task 2-2: Scene B

Work in pairs. Introduce AGM to passengers.

💡 **Useful expressions/patterns**

Greet. / Wonder how to use AGM. / Express thanks.
Explain. / Display steps. / Express good wishes.

💡 **Outlines**

_____
_____

Task 2-3: Scene C

Work in pairs. Introduce Priority seating to the passengers.

💡 **Useful expressions/patterns**

Greet. / Wonder how to use priority seating. / Express thanks.
Explain./ Display steps. / Express good wishes.

💡 **Outlines**

_____
_____
_____

Lesson 7 Station Facilities 车站设备

Name:_____ Class:_____ Date:_____

Self-evaluation

Item	Excellent (90~100分)	Good (80~90分)	Average (60~80分)	Pass (60分)	Fail (<60分)
Be able to use phrases to make conversations					
Take part in pair work and role play					
Improve oral communication skills					
Be able to finish the exercises independently					
Review and preview lessons consciously					

Grade evaluation

Item （项目）	No. （序号）	Criteria （标准）	Score （分数）		
			Self-evaluation （自评）	Peer （学生互评）	Facilitator （指导教师）
General Evaluation （一般评价） （25%）	1	Teamwork, well-assigned jobs（能进行团队合作，能做到分工良好）			
	2	Group discussions for the topic, plan for oral task materials（能进行小组讨论，能准备口语训练任务材料）			
	3	Well-prepared（准备充分）			
	4	Role playing well-performed（角色扮演表现良好）			
	5	Self-confidence（自信）			
Evaluation of Professional Competence （专业能力评价） （25%）	1	Pronunciation, articulation（发音清晰）			
	2	Accuracy, fluency（语言准确、流利）			
	3	Tone of voice, coherent（语调连贯）			
	4	Ability of cross-cultural communication（跨文化交流能力）			
	5	Presentation skills（口语技巧）			
Task-based Evaluation （基于任务的评估） （50%）	1	Describe the pictures（描述图片）			
	2	Practice your oral English and Role play（英语口语练习与角色扮演）			

Lesson 7　Station Facilities 车站设备

Name:_____　Class:_____　Date:_____

Continued

Item （项目）	No. （序号）	Criteria （标准）	Score（分数）		
			Self-evaluation （自评）	Peer （学生互评）	Facilitator （指导教师）
Final Score （合计）		Self-evaluation Score × 20% + Peer Score × 30% + Facilitator Score × 50% （学生自评分数×20%+学生互评分数×30%+教师点评分数×50%）			
Facilitator's Comments （教师评价）					

Task reflection

FACT: What do you get?

FEELING: How do you feel?

FINDING: What do you find?

FUTURE: What shall you do next?

Complementary reading

China's longest subway loop line under construction to pass under Yangtze River

　　Wuhan , as a domestically made shield tunneling machine bored ahead, Line 12 of the Wuhan metro, which will become China's longest subway loop line once completed, began to extend under the Yangtze River on Sunday.

　　The 59.9-km line is the first subway loop line in Wuhan, capital of Central China's

95

Lesson 7   Station Facilities 车站设备

Name:_____   Class:_____   Date:_____

Hubei province, and also the longest nationwide.

Line 12 will connect the three towns of Wuhan, namely Wuchang, Hankou and Hanyang, stretching across seven administrative districts. Meanwhile, the line is designed to go under the Yangtze River twice, the Hanjiang River and two lakes.

The shield tunneling machine used in the project is produced by China Railway Tunnel Group Co., Ltd. and China Railway Engineering Equipment Group Co., Ltd.

The machine weighs about 2,900 tonnes, with an overall length of about 105 meters, and a diameter of 12.56 meters, twice the width of ordinary shield tunneling machines used for subway construction projects.

According to Wuhan Metro Group, the Line 12 will ease the pressure on Wuhan's urban passenger flow upon operation, thus further improving the city's comprehensive transportation system.

Translation:

(1) shield tunneling machine

(2) China's longest subway loop line

(3) the three towns of Wuhan

(4) China Railway Tunnel Group

(5) Wuhan's urban passenger flow

Relevant knowledge

Model Worker Spirit

Lesson 7　Station Facilities 车站设备

Name:_____　Class:_____　Date:_____

In the face of arduous and heavy construction tasks, Liang Xijun (梁西军) has been harvesting and growing all the way, and he has reaped fruitful results on the road of subway construction technology innovation with his unique thickness and tenacity.

Summary

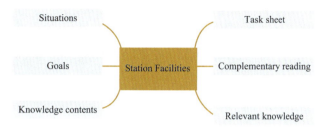

Part Four
Smiling Services

第4部分 微笑服务

Serve people wholeheartedly.
全心全意为人民服务。

Shanghai rail transit staff smile to welcome the World Expo. (Picture from Eastday)
上海轨道交通站务员微笑"迎世博"（图片源自东方网）。

Lesson 8　Customer Service Center
第8课　乘客服务中心
Lesson 9　Service Etiquette
第9课　服务礼仪
Suggested hours: 8 class hours

Lesson 8 Customer Service Center
第8课 乘客服务中心

城市轨道交通客运服务英语口语（第3版）

📚 Overview

Situations (学习情境)

Goals (学习目标)

Groups (任务分组)

Knowledge contents (知识储备)

　　Lead-in (导入)

　　Listening (听力)

　　Conversations (情境对话)

Task sheet (口语训练单)

　　Objectives (训练目标)

　　Tasks (训练内容)

　　Self-evaluation (自我评价)

　　Grade evaluation (成绩评定)

　　Task reflection (任务反思)

Complementary reading (拓展阅读)

Relevant knowledge (知识链接)

Summary (总结)

📚 Situations

- A foreign tourist is asking about something
- A passenger is inquiring something at customer service center
- Metro staffs are helping a passenger

　　Q：作为站务员（Service agent）角色，如何为乘客提供问询服务？如何帮助乘客寻找丢失的物品？

Lesson 8　Customer Service Center 乘客服务中心

Name:_____　Class:_____　Date:_____

 Goals

• Offer inquiry services / consulting services

• Master terminology

• Improve oral communication skills

• Improve professionalism and service awareness

 Groups

Class（班级）		Group No.（小组编号）	
Leader（组长）		Facilitator（指导老师）	
Members（组员）		Roles（角色分工）	

 Knowledge contents

1. Match the words and pictures

Certificate　　　　　　　Phone　　　　　　　Inform

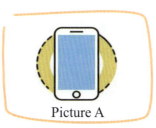
Picture A

Picture B

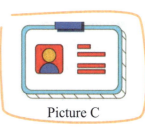
Picture C

2. Match the pictures and sentences

Picture A

(1) It is a piece of equipment that you use to talk to someone else in another place, by dialing a number. It has many functions, such as make a call, play games, navigation, etc.

Picture B

(2) Tell someone about something.

Picture C

(3) It is an official document proving who you are.

101

Lesson 8 Customer Service Center 乘客服务中心

Name:_____ Class:_____ Date:_____

3. Discuss and share

💡 **Do you have any experiences of looking for something or asking for help at the metro station? If so, share with us.**

Listening

1. Listen and tick the topic

Someone wants to rent a car ☐
Linda is going to cinema with her friend ☐
A man is looking for his cellphone at the metro station ☐

2. Listen and complete the blanks

| certificate | shut down | company | contact | inform |
| latest style | trace | phone | model | |

A: Good afternoon, sir. What can I do for you?

B: I have just lost my _____, there are very important _____ numbers in it.

A: Don't worry, sir. I will try my best to help you. Do you remember the last time you used it or saw it, and where was it?

B: I have got a call from my _____ when I just arrived at the metro station.

A: Where did you go after that?

B: I took the metro ticket and went through the AGM.

A: What's your phone number? Has the phone been contacted yet?

B: No, the phone was _____.

A: What is the _____ of your phone? And what's the color?

B: It's a white HUAWEI Mate 40, the _____.

A: May I know your name, and your ID card?

B: OK, my name is Li Zhi, here is my _____.

A: And your contact number. We'll _____ you when we get a _____ of it.

B: Thank you. My phone number is ××××××××××××.

3. Words and expressions

company	['kʌmpənɪ]	n. 公司，商号 v. 结伴，陪伴	trace	[treɪs]	v. 找到；发现；线索；足迹；痕迹
shut down		关机；关闭；停业	contact	['kɒntækt]	n. 联系人；联系方式
latest style		最新款	ID card/identity card		身份证

102

Lesson 8 Customer Service Center 乘客服务中心

Name:_____ Class:_____ Date:_____

4. Read after the recording

Conversations

1. Vocabulary preview

Ornament

It is an attractive object that used to beautify.

There are six keys in total with a small ornament.

Tourist

A person who travels for pleasure.

The Great Wall is a major tourist attraction.

HKD

Hong Kong currency.

The amount of money defrauded in the first 11 months of 2020 alone exceeded HK$560 million ($72.2million).

2. Dialogues

Scene A: A foreign tourist is asking about something

T-Tina (Ticket staff); P-Peter (Passenger)

T: Good morning, sir. May I help you?

P: Good morning. I want to make sure if this is the right line to Hanyang district (汉阳区)?

T: Yes, it is.

P: And I am going to Wangjiawan (王家湾). Which line goes there?

Tina   Peter

T: Line 4. After you pass through the turnstile and go downstairs then you will see indicators which show the directions of the metro. Make sure you are heading the correct direction.

P: Alright. How long does it take then?

T: Within 15 minutes I think.

P: I got it. Thanks for your kindness.

T: You're welcome. Have a good day!

(A sudden meow ...)

T: Excuse me, is that your cat?

P: Yes, my little boy. It is the first time for us to take the Chinese metro.

T: I am afraid you can't take the kitten on the train. According to our regulations, pets are not allowed on the metro train.

P: What a pity! All animals are not allowed?

T: The section 38 of *The Safety Inspection Practice of Wuhan City Rail Transit* stipulates that the Certificate of Vision Impairment and Certificate of Guide Dog shall be provided before people with visual disability travel on the train. And a guide dog shall wear a harness special for guide dogs and the gears preventing it from hurting people. So, I suggest you change to anther transport.

Lesson 8   Customer Service Center 乘客服务中心

Name:_____   Class:_____   Date:_____

P: Thank you all the same.

Scene B: A passenger is inquiring something at customer service center

H-Helen (Service agent); A-Allen (Metro intern); P-Peter (Passenger)

P: I'm sorry to bug you. I am here for sightseeing; I was wondering where I can get an Octopus Card?

Helen   Allen   Peter

A: You can go to customer service center.

P: Customer service center? Where is it?

A: You can find it at every metro station.

P: Okay. Thanks. But one more thing, I can't speak Mandarin Language.

A: Don't worry. The staff in the customer service center can speak Chinese, English and Cantonese.

P: I got it. Thank you again.

(After a few minutes. The foreign tourist now is at the customer service center.)

H: What can I do for you?

P: Hi! I am here for travelling and I want an Octopus Card.

H: Okay. You can rent one.

P: Yeah. That's what I mean. How much is it?

H: Well. It depends on how long you stay here.

P: About a week I guess.

H: In that case I suggest you to rent an ordinary one. The fare in total is HK$ 150. HK$ 50 is for deposit, HK$ 100 is for recharge.

P: Okay, I see. HK$ 150. Here you are.

H: Here is your card. Have fun!

P: Thanks for your time!

H: You're welcome! Good day! Bye!

Scene C: Metro staffs are helping a passenger looking for the missing keys

T-Tina (Ticket staff); H-Helen (Service agent); C-Cherie (Tour guide)

C: Excuse me. Miss, did you see a set of keys?

T: No, I didn't. You can go to the Passenger Service Center, maybe someone has found it.

C: Thank you all the same.

(The passenger goes to the Passenger Service Center.)

H: Good morning ma'am, what can I do for you?

Tina   Helen   Cherie

C: Good morning, I have lost my keys at the metro station.

H: Please let me ask you a few questions concerning your keys. What kind of keys?

C: Six keys and a small ornament.

H: What kind of ornament?

C: It is a pink dolphin ornament that I bought from the Ocean Park in Hong Kong, China.

Lesson 8 Customer Service Center 乘客服务中心

Name:_____   Class:_____   Date:_____

(Claiming for the keys after a moment ...)

H: Is this yours? Our cleaner found it near the toilet.

C: Oh, yes. That's mine. Thank you very much.

H: You are welcome. May I see your identification?

C: Sure, here is my passport.

H: Could you fill in this request slip?

C: OK. Thank you very much.

H: You are welcome.

3. Words and expressions

sightseeing	['saɪtˌsiːɪŋ]	n.	观光；旅游	ordinary	['ɔːdɪnəri]	adj.	普通的
Octopus Card			八达通卡	information	[ˌɪnfə'meɪʃən]	n.	消息；信息
customer	['kʌstəmə(r)]	n.	顾客	indicator	['ɪndɪkeɪtə(r)]	n.	指示牌
service	['sɜːvɪs]	n.	服务	ornament	['ɔːnəmənt]	n.	装饰品
rent	[rent]	v.	租	the Certificate of Vision Impairment			视力损害证明
depend on			取决于	gear	[gɪə(r)]	n.	齿轮；装备

4. Useful sentences and phrases

The same to you.　　你也一样；你也是。

The fare is HK$ 150 in total.　　车费总共是港币150元。

How can I recharge it?　　我怎么充值？

You can recharge it on a self-service device, scan the QR-Code to pay online.　　您可以通过自助设备充值，扫描二维码进行在线支付。

5. Grammar

I am here for sightseeing.

介词for的用法："I am here for sightseeing." "For making more money he got a part-time job."

I want to make sure if this is the right line to Hanyang district.

一般情况下if有"如果""是否"的意思。在本句中"I want to make sure if this is the right line to Hanyang district.", if就是"是否"的意思, 即表达：我想确定一下这条地铁线是否到汉阳区。"I will go swimming if tomorrow is a sunny day."在这句话中if就是"如果"的意思, 即表达：明天如果是晴天我就去游泳。

6. Exercises

Task 1: Fill in the blanks with the words from conversations that match the meanings of sentences

(1) _____ has same meaning as trip.

(2) _____ is made of copper and it's function is to open the door.

(3) _____ means not special.

(4) _____ is just like a guide which tells people directions.

(5) _____ is an ocean creature which has a very smooth skin and very smart.

(6) _____ is a theme park. In that place you can see all types of sea creatures.

Lesson 8 Customer Service Center 乘客服务中心

Name:_____   Class:_____   Date:_____

Task 2: Fill in blanks with the given words

| latest style | change | rent | indicator |
| sightseeing | ornament | in total | |

(1) The _____ shows we should make a right turn after we get through the traffic lights.
(2) I'm going to _____ a bike cycling along the lakeside.
(3) Well ... I don't think this dress is the _____. The color is so old-fashioned.
(4) That would be 200 RMB _____.
(5) The _____ is quite spectacular on the top of mountain.
(6) Coco has bought a small _____ as a souvenir.

Task 3: Translate the following sentences into English

（1）上周末琳达跟她的朋友坐地铁去了海洋公园。

（2）根据规定，禁止携带宠物上车（地铁）。

（3）从火车南站到机场一共有21站。

（4）这个小装饰品是我去年在新加坡旅游的时候买的。

（5）他在乘客服务中心找到了她的一串钥匙。

Task sheet

Objectives

Learn how to describe the duty of service agent.
Learn how to offer help to passengers.

Oral task of grading details

Specific requirements for the academic quality of College English in higher Vocational Education.

Levels（水平分类）	Qualitative description（质量描述）
Level 1 (general requirements) ［水平一（一般要求）］	Level 1-1: Can basically understand clear pronunciation, slow speech in daily life and urban rail transit passenger service posts related topics［能基本听懂发音清晰、语速较慢的日常生活语篇和职场（城市轨道交通客运服务岗位）相关话题］ Level 1-2: Can basically understand the relevant English materials of urban rail transit passenger transport service, understand the main contents and obtain key information; understand the cultural connotation and identify the professional terms of urban rail transit passenger transport service（能基本读懂、看懂城市轨道交通客运服务的相关英语资料，理解主要内容，能获取关键信息；领会文化内涵，能识别城市轨道交通客运服务的专业术语）

Lesson 8   Customer Service Center 乘客服务中心

Name:_____   Class:_____   Date:_____

Continued

Levels（水平分类）	Qualitative description（质量描述）
Level 1 (general requirements) ［水平一（一般要求）］	Level 1-3: Can communicate with others on familiar topics in daily life and urban rail transit passenger transport service; the expression is basically accurate and fluent, and can briefly introduce workplace culture and METRO culture（能在日常生活中和城市轨道交通客运服务工作中就比较熟悉的话题与他人进行语言交流，表达基本准确、流畅；能简单介绍职场文化和地铁企业文化）
	Level 1-4: Can briefly express their experiences, opinions and feelings; the sentences are basically correct and expressed clearly（能简要表达自己的经历、观点、感受；语句基本正确，表达清楚，格式恰当）
	Level 1-5: Can meet the basic communication needs on familiar with daily life and urban rail transit passenger transport service（能就日常生活和城市轨道交通客运服务工作中熟悉的话题，满足基本沟通需求）
	Level 1-6: Be able to make a clear learning plan; to obtain learning resources through online and offline channels under the guidance of teachers（能制订明确的学习计划；能在教师引导下通过线上线下多种渠道获取学习资源）

Tasks

Task 1. Describe the pictures
Task 1: Service agent

(Source:Hubei Publishing, 2022-5-7)

💡 **Way out**

How many people in the picture? Are they working? What do they do?

💡 **Language notes**

customer service center   乘客服务中心
在城市轨道交通系统内设置的为乘客提供票务、咨询等客运服务或延伸服务的场所。
service Signs   服务标志
通过颜色、图形或文字的组合，表达客运服务信息的设施。
专业术语来源：《城市轨道交通工程基本术语标准》（GB/T 50833—2012）
（*Standard for basic terminology of urban rail transit engineering*）

💡 **Notes**

城市轨道交通客运服务，年度统计数据满足以下指标要求：

107

Lesson 8　Customer Service Center 乘客服务中心

Name:_____　　Class:_____　　Date:_____

列车正点率应大于或等于98.5%；

列车运行图兑现率应大于或等于99%；

有效乘客投诉率不应超过3次/百万人次，有效乘客投诉回复率应为100%。

资料来源：《城市轨道交通运营管理规范》（GB/T 30012—2013）（*Regulations for operation management of urban rail transit*）

💡 **Outlines**

_____

_____

_____

Task 2. Practice your oral English and Role play

Task 2-1: Scene A

　　Work in pairs. Suppose you were a staff of customer service center and someone is asking you about something...

💡 **Useful expressions/patterns**

　　Good morning / afternoon. May I help you? / What can I do for you? / Is there anything I can do for you...

　　Yes. How can I rent a card.../ I have no idea what's wrong with my token...

💡 **Outlines**

_____

_____

_____

Task 2-2: Scene B

Work in pairs. Suppose you were a passenger, and now you are at metro customer service center...

💡 **Useful expressions/patterns**

　　Excuse me. I think I lost my phone / watch at this station. Could you please help me to find it?

　　Sure / Of course / Certainly...Can you tell me when and where did you last see your phone/ watch? What's style / color of your phone / watch?

💡 **Outlines**

_____

_____

_____

Lesson 8 Customer Service Center 乘客服务中心

Name:_____   Class:_____   Date:_____

Self-evaluation

Item	Excellent (90~100分)	Good (80~90分)	Average (60~80分)	Pass (60分)	Fail (<60分)
Be able to use phrases to make conversations					
Take part in pair work and role play					
Improve oral communication skills					
Be able to finish the exercises independently					
Review and preview lessons consciously					

Grade evaluation

Item (项目)	No. (序号)	Criteria (标准)	Score (分数)		
			Self-evaluation (自评)	Peer (学生互评)	Facilitator (指导教师)
General Evaluation (一般评价) (25%)	1	Teamwork, well-assigned jobs（能进行团队合作，能做到分工良好）			
	2	Group discussions for the topic, plan for oral task materials（能进行小组讨论，能准备口语训练任务材料）			
	3	Well-prepared（准备充分）			
	4	Role playing well-performed（角色扮演表现良好）			
	5	Self-confidence（自信）			
Evaluation of Professional Competence (专业能力评价) (25%)	1	Pronunciation, articulation（发音清晰）			
	2	Accuracy, fluency（语言准确、流利）			
	3	Tone of voice, coherent（语调连贯）			
	4	Ability of cross-cultural communication（跨文化交流能力）			
	5	Presentation skills（口语技巧）			
Task-based Evaluation (基于任务的评估) (50%)	1	Describe the pictures（描述图片）			
	2	Practice your oral English and Role play（英语口语练习与角色扮演）			

Lesson 8　Customer Service Center 乘客服务中心

Name:_____　Class:_____　Date:_____

Continued

Item （项目）	No. （序号）	Criteria （标准）	Score（分数）		
			Self-evaluation （自评）	Peer （学生互评）	Facilitator （指导教师）
Final Score （合计）		Self-evaluation Score × 20% + Peer Score × 30% + Facilitator Score × 50% （学生自评分数×20%+学生互评分数×30%+教师点评分数×50%）			
Facilitator's Comments （教师评价）					

Task reflection

FACT: What do you get?

FEELING: How do you feel?

FINDING: What do you find?

FUTURE: What shall you do next?

Complementary reading

Station services in MTR

Free internet service:

You can now enjoy free Wi-Fi in every MTR station! You can connect to the internet with your smartphone or computer with internet accessibility near the sign of *MTR Free Wi-Fi Hotspots*. Simply select the *MTR Free Wi-Fi* network for free connectivity of up to 15 minutes per session with maximum 5 sessions for each smart phone/computer per day. So you can stay connected wherever you go!

Free publications:

Short of something interesting to read on the train? The following four free publications are distributed in all MTR stations on the Kwun Tong Line (观塘线), Tsuen Wan Line (荃湾线), Island Line (港岛线), Tung Chung Line (东涌线) and Tseung Kwan O Line (将军澳线):

　Metro Daily　daily newspaper (Mondays—Fridays, except public holidays)

　Metro Pop　infotainment magazine (Thursdays, except public holidays)

　Job Market　recruitment magazine (Tuesdays & Fridays, except public holidays) distributed at Kwun Tong Line, Tsuen Wan Line, Island Line, Tseung Kwan O Line and

Lesson 8 Customer Service Center 乘客服务中心

Name:_____ Class:_____ Date:_____

Tung Chung Line (Tsing Yi and Tung Chung stations)

Recruit recruitment magazine (Tuesdays & Fridays, except public holidays) distributed at Tung Chung Line (Hong Kong and Olympic stations)

ATM service:

With Automated Teller Machines (ATMs) installed at almost all stations, the MTR also makes it incredibly easy to obtain cash any time you need it.

Post boxes:

If you need to send mail, simply drop your letters in the conveniently located Post boxes at the stations across Hong Kong. Post box services are available at the following stations.

Translation:

(1) You can now enjoy free Wi-Fi in every MTR station!

(2) So you can stay connected wherever you go!

(3) Short of something interesting to read on the train?

(4) The MTR also makes it incredibly easy to obtain cash any time you need it.

(5) If you need to send mail, simply drop your letters in the conveniently located Post boxes at the stations across Hong Kong.

Relevant knowledge

Model Worker Spirit

Lesson 8　Customer Service Center 乘客服务中心

Name:_____　　Class:_____　　Date:_____

I will continue to make contributions in my post and tell the story of more than 30,000 of our Shanghai Metro people and the story of the subway, said Gao Yu.

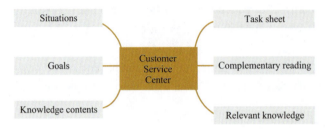

Lesson 9 Service Etiquette
第9课 服务礼仪

 **Overview**

Situations (学习情境)

Goals (学习目标)

Groups (任务分组)

Knowledge contents (知识储备)

    Lead-in (导入)

    Listening (听力)

    Conversations (情境对话)

Task sheet (口语训练单)

    Objectives (训练目标)

    Tasks (训练内容)

    Self-evaluation (自我评价)

    Grade evaluation (成绩评定)

    Task reflection (任务反思)

Complementary reading (拓展阅读)

Relevant knowledge (知识链接)

Summary (总结)

 Situations

- Platform service etiquette
- Rider etiquette
- Public etiquette

    Q：刚刚入职的地铁公司实习生会（Metro intern）学习到哪些礼仪呢？

113

Lesson 9　Service Etiquette 服务礼仪

Name:_____　　Class:_____　　Date:_____

 Goals

- Understand Service etiquette
- Master terminology
- Improve oral communication skills
- Improve professionalism and service awareness

 Groups

Class（班级）		Group No.（小组编号）	
Leader（组长）		Facilitator（指导老师）	
Members（组员）		Roles（角色分工）	

Knowledge contents

Lead-in

1. Match the words and pictures

Etiquette　　　　　　　Cheerful　　　　　　　Supervisor

Picture A

Picture B

Picture C

2. Match the pictures and sentences

Picture A

(1) The rules indicating the proper and polite way to behave.

Picture B

(2) Feeling or showing that you are willing to do something without complaining, feeling or showing happiness.

Picture C

(3) An administrative officer in charge of a business, government, or school unit or operation.

Lesson 9　Service Etiquette 服务礼仪

Name:_____　Class:_____　Date:_____

3. Discuss and share

💡 Do you know something about service etiquette? If you do, tell us about it.

Listening

1. Listen and tick the topic

Service etiquette tips ☐
Guests first ☐
Customer service ☐

2. Listen and complete the blanks

| essential | situation | standards | tips | smile | basic |
| first name | empathize | fair | manner | cheerful | |

Service etiquette is the _____ quality and _____ requirement for customer service.

We should pay attention to appearance, _____ and language, comply with the _____ of the customer service, provide considerate service, so as to show the good professional accomplishment.

Here are some service etiquette _____.

Be Cheerful

_____ is the most beautiful language. Smile service for guests, can create a relaxed and cordial atmosphere. A smile and _____ demeanor can make guests feel at home.

Be Professional

Refer to customers as sir and madam. Refrain from using slang words or dirty words, and do not refer to the customer by the _____ unless he insists that you do so. All of these help to set a professional etiquette service for your customers.

Use Honesty

Tell the truth to the customer at all times. Let him/her know what you are authorized to do to help in a situation, and if the _____ gets beyond your authority then let the customer know that you need to transfer to a supervisor.

Be Empathetic

Listen to what the customer has to say and then let him know that you understand his situation. _____ with the customer by telling him how you would feel if you were in his situation. This will help the customer to believe you are on his side, and that you are going to work to offer a _____ solution.

3. Words and expressions

comply with			遵从	refer to		描述；提到
considerate	[kən'sɪdərət]	adj.	考虑周到的；体贴的	slang words		俚语

Lesson 9 Service Etiquette 服务礼仪

Name:_____ Class:_____ Date:_____

accomplishment	[əˈkʌmplɪʃmənt]	n.	成就，成绩；才艺	at all times			总是，一直；随时
cordial	[ˈkɔːdjəl]	adj.	友好的，可亲的	empathetic	[ˌempəˈθetɪk]	adj.	移情作用的，感情移入的

4. Read after the recording

Conversations

1. Vocabulary preview

Uniform

Dress of a distinctive design or fashion worn by members of a particular group and serving as a means of identification.

He looks quite smart in his uniform.

Platform

It is the area beside the tracks where you wait for or get off a train in the station.

The train was about to leave and I was not even on the platform.

Don't

A command or entreaty not to do something.

Here is a list of don'ts.

2. Dialogues

Scene A: Platform service etiquette

A-Allen (Metro intern); T-Tina (Ticket staff)

T: How about your first day working in the metro station?

A: I was a little nervous when I put on my uniform and stood on the platform.

T: Are you working on the platform?

A: Yes. Platform is one of the key positions of metro station for passenger service. The passenger flow on the platform is relatively large, so the platform service should combine safety with service etiquette.

Allen   Tina

T: What do you feel about the service etiquette?

A: First of all, uniform is needed. The appearance should be neat and dignified, which should meet the requirements of *Urban rail passenger transport service*. Do not wear slippers and leave long nails. Then stand upright, and hands are naturally vertical; do not cross hands in front of the chest or put them into pockets.

T: What do you do when the passengers get on and get off the train?

A: My duties are to maintain good platform order, remind the passenger to stand at the yellow line to queue up for train, arrange them to wait on either side and yield to alighting passengers, and noting the gaps between the platforms.

T: Not easy at work! Are you satisfied with it?

A: Although tired and humble, it is my honor to serve our passengers!

Lesson 9　Service Etiquette 服务礼仪

Name:_____　Class:_____　Date:_____

Scene B: Rider etiquette

A-Allen (Metro intern); H-Helen (Service agent)

A: Do you know the rider etiquette while taking subway?

H: Of course! First let passengers exit the train before attempting to board, especially during rush hours. This helps speed up service. Second, offer your seat to seniors, pregnant women, and people with disabilities.

A: What are the other don'ts?

Allen　　Helen

H: There are as followings: don't eat or drink on the train; don't have loud personal conversations and use mobile or music devices without headphones; don't leave litter ...

A: Are scooters allowed in the station or on the subway?

H: According to applicable regulations, scooters are not banned from the stations, but according to section 43 of *Safe Operation Rules of Beijing Rail Transit*, chasing, fighting, scootering, roller skating, and bicycling are not allowed in the station and on the subway. We staff has the right to stop that action.

A: It seems that you are familiar with the etiquette and tips of the subway.

H: That's for sure! Good manners go a long way toward a smooth transit experience.

A: Follow these etiquette and tips for riders to have a pleasant journey. We all have to keep them.

H: Absolutely!

Scene C: Public etiquette

A-Allen (Metro intern); H-Helen (Service agent)

A: In addition to rider etiquette, do you know what public etiquette needs to be strictly observed?

O: There are many regulations. Simply speaking, no smoking in the public and no use of dirty words; no litter dropping or spitting in public areas; if you take the escalator, keep right to leave space for others to pass by; don't speak loudly or make noise in public areas; wait in line, never jump the queue; don't damage public property ...

Allen　　Helen

3. Words and expressions

uniform	['juːnɪfɔːm]	n.	制服	passenger flow		客流，客流量
platform	['plætfɔːm]	n.	站台	combine ... with		与……结合（化合）
upright	['ʌpraɪt]	adj.	笔直的	be familiar with		熟悉
humble	['hʌmbl]	adj.	谦逊的；简陋的	headphone	['hedfəʊn] n.	耳机；听筒
remind	[rɪ'maɪnd]	n.	提醒	It is my honor to ...		这是我的荣幸……
yield to			让步			
Urban rail passenger transport service				《城市轨道交通客运服务》（GB/T 22486—2008）		
Safe Operation Rules of Beijing Rail Transit				《北京轨道交通安全操作规程》		

117

Lesson 9   Service Etiquette 服务礼仪

Name:_____   Class:_____   Date:_____

4. Useful sentences and phrases

This helps speed up service.   这有助于加快(提升)服务。
That's for sure!   那是肯定的！
Simply speaking ...   简单地说……
Good manners go a long way toward a smooth transit experience.   良好的礼仪对顺畅的交通体验大有帮助。
Wait in line, never jump the queue.   排队等候，千万不要插队。

5. Grammar

询问对方意见的表达方式：
How about your first day working in the metro station?
What do you feel about the service etiquette?
What do you do when the passengers get on and get off the train?
Are you satisfied with it?
Do you know the rider etiquette while taking subway?

6. Exercises

Task 1: Fill in the blanks with the words from conversations that match the meanings of sentences

(1) _____ is the clothing of distinctive design worn by members of a particular group as a means of identification.

(2) If you are sitting or standing _____, you are sitting or standing with your back straight, rather than bending or lying down.

(3) If you _____ two or more things, they join together to make a single thing.

(4) If someone or something is _____ to you, you recognize them or know them well.

(5) _____: is held over or inserted into the ear; elector-acoustic transducer for converting electric signals into sounds.

Task 2: Fill in the blanks with the given words

| uniform | yield to | cheerful | humble | headphone | honor | upright |

(1) A _____ person is not proud and does not believe that they are better than other people.

(2) She's already outgrown her school _____.

(3) She has a _____ outlook on life.

(4) He sat _____ in his chair.

(5) They _____ the wind, but they never break.

(6) What is Dolby _____ Technology?

(7) It's my great _____ to attend the award ceremony today.

Task 3: Translate the following sentences into English

（1）主管在他酒后上班时解雇了他。

Lesson 9　Service Etiquette 服务礼仪

Name:_____　　Class:_____　　Date:_____

（2）不要把双手交叉在前胸。

（3）不要穿拖鞋。

（4）你熟悉这个电脑新软件吗？

（5）请站在黄线处排队等车。

（6）请在两侧候车，先下后上。

（7）当乘坐自动扶梯时，请将右边的位置让给需要通过的人。

Task sheet

Objectives

Learn how to give etiquette tips.
Learn how to standardize the behavior of taking the subway/metro.

Oral task of grading details

Specific requirements for the academic quality of College English in higher Vocational Education.

Levels（水平分类）	Qualitative description（质量描述）
Level 1 (general requirements) ［水平一（一般要求）］	Level 1-1: Can basically understand clear pronunciation, slow speech in daily life and urban rail transit passenger service posts related topics［能基本听懂发音清晰、语速较慢的日常生活语篇和职场（城市轨道交通客运服务岗位）相关话题］
	Level 1-2: Can basically understand the relevant English materials of urban rail transit passenger transport service, understand the main contents and obtain key information; understand the cultural connotation and identify the professional terms of urban rail transit passenger transport service（能基本读懂、看懂城市轨道交通客运服务的相关英语资料，理解主要内容，能获取关键信息；领会文化内涵，能识别城市轨道交通客运服务的专业术语）
	Level 1-3: Can communicate with others on familiar topics in daily life and urban rail transit passenger transport service; the expression is basically accurate and fluent, and can briefly introduce workplace culture and METRO culture（能在日常生活中和城市轨道交通客运服务工作中就比较熟悉的话题与他人进行语言交流，表达基本准确、流畅；能简单介绍职场文化和地铁企业文化）
	Level 1-4: Can briefly express their experiences, opinions and feelings; the sentences are basically correct and expressed clearly（能简要表达自己的经历、观点、感受；语句基本正确，表达清楚，格式恰当）
	Level 1-5: Can meet the basic communication needs on familiar with daily life and urban rail transit passenger transport service（能就日常生活和城市轨道交通客运服务工作中熟悉的话题，满足基本沟通需求）
	Level 1-6: Be able to make a clear learning plan; to obtain learning resources through online and offline channels under the guidance of teachers（能制订明确的学习计划；能在教师引导下通过线上线下多种渠道获取学习资源）

Lesson 9　Service Etiquette 服务礼仪

Name:_____　　Class:_____　　Date:_____

Tasks

Task 1. Describe the pictures

Task 1-1: Photo Etiquette

💡 **Way out**

What is phone etiquette?
Try to describe the details.

💡 **Language notes**

platform　站台
车站内供乘客候车和乘降的平台。

trip　出行
从出发地到目的地的交通行为。

专业术语来源：《城市轨道交通工程基本术语标准》（GB/T 50833—2012）
（Standard for basic terminology of urban rail transit engineering）

💡 **Outlines**

_____
_____
_____

Task 1-2: Social Media Etiquette Tips

120

Lesson 9　Service Etiquette 服务礼仪

Name:_____　Class:_____　Date:_____

💡 Way out

What is social media etiquette?
Try to describe the details.

💡 Outlines

_____

_____

Task 1-3: Live Chat Etiquette Tips

💡 Way out

What is live chat etiquette?
Try to describe the details.

💡 Notes

客运组织服务

运营单位应加强服务管理，改进和提高客运服务质量，并应采取以下措施：

加强员工培训，增强爱岗敬业和优质服务意识；

提高员工的规范服务技能和业务水平；

建立与乘客沟通渠道，加强与乘客沟通；

建立投诉监督机制，接受社会监督。

资源来源：《城市轨道交通运营管理规范》（GB/T 30012—2013）（*Regulations for operation management of urban rail transit*）

💡 Outlines

_____

_____

_____

121

Lesson 9　Service Etiquette 服务礼仪

Name:_____　Class:_____　Date:_____

Task 2. Practice your oral English and Role play

Task 2-1: Scene A

Work in pairs. Suppose someone were naked to the waist in the metro station.

💡 **Useful expressions/patterns**

Greet. /Show the regulations and etiquette...

Explain. / Give reasons for that...

💡 **Outlines**

_____
_____
_____

Task 2-2: Scene B

Work in pairs. Suppose someone were spitting on the train.

💡 **Useful expressions/patterns**

Greet. /Show the regulations and etiquette...

Explain. / Give reasons for that...

💡 **Outlines**

_____
_____
_____

Self-evaluation

Item	Excellent (90~100分)	Good (80~90分)	Average (60~80分)	Pass (60分)	Fail (<60分)
Be able to use phrases to make conversations					
Take part in pair work and role play					
Improve oral communication skills					
Be able to finish the exercises independently					
Review and preview lessons consciously					

Lesson 9   Service Etiquette 服务礼仪

Name:_____ Class:_____ Date:_____

Grade evaluation

Item（项目）	No.（序号）	Criteria（标准）	Score（分数）		
			Self-evaluation（自评）	Peer（学生互评）	Facilitator（指导教师）
General Evaluation（一般评价）（25%）	1	Teamwork, well-assigned jobs（能进行团队合作，能做到分工良好）			
	2	Group discussions for the topic, plan for oral task materials（能进行小组讨论，能准备口语训练任务材料）			
	3	Well-prepared（准备充分）			
	4	Role playing well-performed（角色扮演表现良好）			
	5	Self-confidence（自信）			
Evaluation of Professional Competence（专业能力评价）（25%）	1	Pronunciation, articulation（发音清晰）			
	2	Accuracy, fluency（语言准确、流利）			
	3	Tone of voice, coherent（语调连贯）			
	4	Ability of cross-cultural communication（跨文化交流能力）			
	5	Presentation skills（口语技巧）			
Task-based Evaluation（基于任务的评估）（50%）	1	Describe the pictures（描述图片）			
	2	Practice your oral English and Role play（英语口语练习与角色扮演）			
Final Score（合计）		Self-evaluation Score × 20% + Peer Score × 30% + Facilitator Score × 50%（学生自评分数×20%+学生互评分数×30%+教师点评分数×50%）			
Facilitator's Comments（教师评价）					

Task reflection

FACT: What do you get?

FEELING: How do you feel?

FINDING: What do you find?

FUTURE: What shall you do next?

Lesson 9　Service Etiquette 服务礼仪

Name:_____　Class:_____　Date:_____

📖 Complementary reading

New rules promise better subway etiquette

The new regulation of dos and don'ts for people using urban rail transit services, which was issued by the Ministry of Transport recently, may seem a no-brainer for anyone who is aware of some basic social etiquette. But its publication—and implementation starting April 1 next year—is timely and necessary given the rising number of complaints about uncivilized and sometimes unruly behavior on the subway and light rail services and the inefficiency in dealing with such problems because of the lack of national unified norms.

Now the national regulation explicitly puts such behavior as eating, littering, urinating, and graffitiing on the must-not-do list for passengers. Other offenses include smoking, forcing open train doors, lying on seats, making loud noises and using electronic instruments without headphones. Violators will be dealt with by relevant departments according to the law, the new regulation states, although no specific penalties have been listed, obviously giving local authorities free reign to work out their own punishments based on their actual conditions.

Translation:

(1) The new regulation of dos and don'ts for people using urban rail transit services, which was issued by the Ministry of Transport recently.

(2) Unruly behavior on the subway and light rail services.

(3) They sometimes are too simplistic or even contradictory making the rules hard to enforce.

(4) Now the national regulation explicitly puts such behavior as eating, littering, urinating, and graffitiing on the must-not-do list for passengers.

(5) Other offenses include smoking, forcing open train doors, lying on seats, making loud noises and using electronic instruments without headphones.

📖 Relevant knowledge

Service etiquette exercise of Tianjin Metro

Standardized movements are used to show the service image of Tianjin Metro. All employees must carry out the standard training of this set of civilized service etiquette

Lesson 9　Service Etiquette 服务礼仪

Name:_____　Class:_____　Date:_____

before taking up their posts, so that this set of exercises will become full and normal. In normal times, all employees should often practice and standardize service etiquette, and let passengers enjoy a high level of etiquette service when taking the subway.

Summary

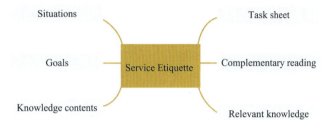

Part Five
Emergency

第5部分 紧急情况

Safety is no small matter, responsibility is greater than anything.
安全无小事,责任重于山。

"We will promote safe development, and raise public awareness that life matters most and that safety comes first."
"我们将促进安全发展,并提高公众对生命最重要、安全第一的认识。"

Lesson 10　Emergency Handling
第10课　应急处理
Suggested hours: 4 class hours

Lesson 10　Emergency Handling
第10课　应急处理

城市轨道交通客运服务英语口语（第3版）

Overview

Situations (学习情境)

Goals (学习目标)

Groups (任务分组)

Knowledge contents (知识储备)

　　Lead-in (导入)

　　Listening (听力)

　　Conversations (情境对话)

Task sheet (口语训练单)

　　Objectives (训练目标)

　　Tasks (训练内容)

　　Self-evaluation (自我评价)

　　Grade evaluation (成绩评定)

　　Task reflection (任务反思)

Complementary reading (拓展阅读)

Relevant knowledge (知识链接)

Summary (总结)

Situations

- An emergency at the station
- An emergency on the train
- A fire drill

　　Q：假如你是一名站务员（Service agent）或地铁公司实习生（Metro intern），你会如何面对地铁车站内乘客的心脏病突发情况？如何组织地铁车站的消防演习？如何处理地铁火灾的情况？

Lesson 10　Emergency Handling 应急处理

Name:_____　　Class:_____　　Date:_____

 Goals

- Deal with emergencies
- Master terminology
- Improve oral communication skills
- Improve professionalism and service awareness

 Groups

Class（班级）		Group No.（小组编号）	
Leader（组长）		Facilitator（指导老师）	
Members（组员）		Roles（角色分工）	

 Knowledge contents

Lead-in

1. Match the words and pictures

System　　　　　　　　Child　　　　　　　　Write

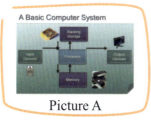
Picture A

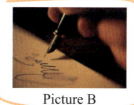
Picture B

Picture C

2. Match the pictures and sentences

Picture A

(1) It is a set of devices powered by electricity, for example a computer or an alarm.

Picture B

(2) A human offspring of any age, son or daughter, kid.

Picture C

(3) You record something on the notebook with your hand and pen.

129

Lesson 10　Emergency Handling 应急处理

Name:_____　Class:_____　Date:_____

3. Discuss and share

💡 Have you ever encountered an emergency? If you have, tell us something about it.

Listening

1. Listen and tick the topic

This conversation took place in a hospital　☐
The woman is worried about her missing child　☐
The house is on fire　☐

2. Listen and complete the blanks

| Lily Parker | ten | locate | madam | personal particulars |
| broadcasting | six | red dress | contact | Exit A |

A: Good afternoon, _____. What can I do for you?

B: I have lost my daughter _____ somewhere around the _____. Please help me to find her ...

A: I can understand your feelings, don't worry, madam. How old is she?

B: She is about _____.

A: What clothes did she wear?

B: She was wearing a _____ this morning.

A: And when did you see her last?

B: About _____ minutes ago.

A: Please write down your daughter's _____ and we would try our best to _____ your child over our system.

B: Thank you very much.

Broadcasting:

Ladies and gentlemen, may I have your attention please? We are looking for Lily Parker, a six-year-old girl in a red dress. Her mother is waiting for her in the station_____ room. If you find her, please _____ us.

3. Words and expressions

| broadcasting room | | 广播站 | personal particulars | | 个人情况，详情 |
| I understand how you feel | | 感同身受 | system | ['sɪstəm] | 系统 |

4. Read after the recording

Conversations

1. Vocabulary preview

Aspirin
A tablet used medicinally to reduce fever and relieve pain.

Lesson 10 Emergency Handling 应急处理

Name:_____ Class:_____ Date:_____

Take an aspirin and lie down.

Emergency

A serious, unexpected and dangerous situation requiring immediate action.

An emergency exit.

Immediately

Right away, instantly, at once, without delay or hesitation.

The pain immediately eased.

2. Dialogues

Scene A: An emergency at the station

H-Helen (Service agent); M-Mr. Thomas (Passenger); P-Peter(Passenger)

H: Be careful! Sir, please avoid the gap between the edge door and the train, it is dangerous.

M: Oh, thank you! I am not feeling well. Could you get a doctor for me?

H: Sure, I will inform our resident doctor right away. Please wait a moment.

Helen

Thomas

Peter

M: Yes, please.

P: Good afternoon, sir. I am doctor. What's the matter with you?

M: I've got a sore throat. And a little sick!

P: Are you cold?

M: Yes, I am shivering, and I coughed a lot at night.

P: How long have you been like this?

M: About three days.

P: You are running a temperature. Open your mouth and say "Ah".

M: Ah ...

P: Let me examine your chest. I think you have got the flu. Please take one tablet three times a day after meals.

M: What is the medicine for?

P: The aspirins will relieve your body.

M: One tablet three times a day after meals. I got it.

P: You'd better stay in bed. You will get better soon.

M: I'll take your advice. Thank you.

Scene B: An emergency on the train

A-Allen (Metro intern); P-Peter (Passenger)

A: What's wrong with you, madam? You look pale.

P: I am suffering a heart trouble and my blood pressure is a little high.

A: Really? But we don't have a doctor on the train.

P: Oh, my god, I feel like vomiting.

A: Sit down, please. You can have some water to make you feel better.

P: Thank you. I have some medicines for heart disease.

A: Where are the medicines?

Allen

Lesson 10　Emergency Handling 应急处理

Name:_____　Class:_____　Date:_____

P: Just in my coat pocket.
A: Here you are. Take the medicines quickly.
P: Thank you.

Scene C: A fire drill

H-Helen (Service agent); A-Allen (Metro intern); P-Peter (Passenger)

Broadcasting:

May I have your attention, please? This is an emergency. The place is on fire now. Please follow the instruction of station staff and leave the station in order. Withdraw this building immediately. Thank you for your cooperation, this is a fire drill.

　　　　　　Helen　　　　Peter　　　　Allen

H: Helen speaking, may I help you?
P: I am trapped in the metro station. What shall I do?
H: Don't worry! It is a fire drill.
P: Oh, could you send someone to help me?
H: What's your exact position?
P: I am in the front of the ticket machine.
H: Could you stay where you are? We will send someone immediately.
P: OK, please be quick!
(After 5 minutes ...)
A: Hello, is anyone there?
P: Yes, I am here.
A: Could you step to the fire escape and follow the emergency exit door.
P: Sure.

3. Words and expressions

gap	[gæp]	n.	空隙	fire drill		消防训练
sick	[sɪk]	adj.	恶心的，不舒服的	running a temperature		测体温
vomit	['vɒmɪt]	v.	呕吐	blood pressure		血压
trap	[træp]	v.	诱骗，困住			

4. Useful sentences and phrases

take a seat　　坐下
not feel well　　身体不适
How long have you been like this?　你这样多久了？
stay warm/keep warm　　保暖
in case to　　以防万一

5. Grammar

用英语描述病情：
He feels headache and vomiting.　　他感觉头疼和想吐。

He feels light-headed.　他感觉头晕。
He has been lacking in energy for some time.　他感到虚弱有一段时间了。
He has a persistent cough.　他在不停地咳嗽。
You should wear a mask when you catch a cold.　你感冒的时候要戴好口罩。
I have a bloated, uncomfortable feeling after meal.　我饭后肚子胀，不舒服。
He has pain on the sole of his feet.　他脚底很痛。
He is sleeping poorly.　他睡不好。
My blood pressure is really up.　我血压很高。
His breathing has become increasingly difficult.　他呼吸越来越困难。
I have some problems with my teeth.　我的牙齿有问题。
Her eyes seem to be bulging.　她的眼睛有点肿胀。
He has a repeated buzzing or other noises in his ears.　他耳朵有嗡嗡声。

6. Exercises

Task 1: Fill in the blanks with the words from conversations that match the meanings of sentences

(1) A _____ is a space between two things or a hole in the middle of something solid.

(2) If someone looks _____, their face looks a lighter color than usual, usually because they are ill, frightened, or shocked.

(3) The _____ in a place or container is the force produced by the quantity of gas or liquid in that place or container.

(4) If you _____, food and drink comes back up from your stomach and out through your mouth.

(5) _____ is a substance that you drink or swallow in order to cure an illness.

(6) If you are _____, you are ill.

(7) When groups of people _____, they leave the place where they are fighting or where they are based and return nearer home.

Task 2: Fill in the blanks with the given words

> suffering　　edge door　　examine　　relieve　　heart disease

(1) Please avoid the gap between the _____ and the train.

(2) I am _____ a heart trouble.

(3) I have some medicines for _____.

(4) Let me _____ your chest. I think you have got the flu.

(5) The aspirins will _____ your body.

Task 3: Translate the following sentences into English

（1）我马上通知车站医生。

（2）我的血压有点高。

Lesson 10　Emergency Handling 应急处理

Name:_____　　Class:_____　　Date:_____

（3）你喝点水会让你感觉好些。

（4）给你（药），赶紧吃药。

（5）一天服三次药，每次一颗，饭后吃。

 Task sheet

Objectives

Learn how to describe the symptoms of illness.
Learn how to deal with emergency.

Oral task of grading details

Specific requirements for the academic quality of College English in higher Vocational Education.

Levels（水平分类）	Qualitative description（质量描述）
Level 1 (general requirements)［水平一（一般要求）］	Level 1-1: Can basically understand clear pronunciation, slow speech in daily life and urban rail transit passenger service posts related topics［能基本听懂发音清晰、语速较慢的日常生活语篇和职场（城市轨道交通客运服务岗位）相关话题］
	Level 1-2: Can basically understand the relevant English materials of urban rail transit passenger transport service, understand the main contents and obtain key information; understand the cultural connotation and identify the professional terms of urban rail transit passenger transport service（能基本读懂、看懂城市轨道交通客运服务的相关英语资料，理解主要内容，能获取关键信息；领会文化内涵，能识别城市轨道交通客运服务的专业术语）
	Level 1-3: Can communicate with others on familiar topics in daily life and urban rail transit passenger transport service; the expression is basically accurate and fluent, and can briefly introduce workplace culture and METRO culture（能在日常生活中和城市轨道交通客运服务工作中就比较熟悉的话题与他人进行语言交流，表达基本准确、流畅；能简单介绍职场文化和地铁企业文化）
	Level 1-4: Can briefly express their experiences, opinions and feelings; the sentences are basically correct and expressed clearly（能简要表达自己的经历、观点、感受；语句基本正确，表达清楚，格式恰当）
	Level 1-5: Can meet the basic communication needs on familiar with daily life and urban rail transit passenger transport service（能就日常生活和城市轨道交通客运服务工作中熟悉的话题，满足基本沟通需求）
	Level 1-6: Be able to make a clear learning plan; to obtain learning resources through online and offline channels under the guidance of teachers（能制订明确的学习计划；能在教师引导下通过线上线下多种渠道获取学习资源）

Lesson 10　Emergency Handling 应急处理

Name:_____　Class:_____　Date:_____

Tasks

Task 1. Describe the pictures

Task 1-1: Symptoms of illness

 Way out

What's the matter?

Describe her symptoms: sneezing/runny nose...

How to deal with? Take the medicine? See the doctor? Wear a mask? Take an isolated observation?

💡 **Language notes**

fire alarm system　　　火灾自动报警系统

实现火灾监测、自动报警并直接联动消防救灾设备的自动控制系统。

platform screen door　　站台（屏蔽）门

设置在站台边缘，将乘客候车区与列车运行区相互隔离，并与列车门相对应、可多级控制开启与关闭滑动门的连续屏障，有全高、半高、密闭和非密闭之分。

emergency escape door　　应急门

当列车门与滑动门不能对齐时，供疏散的门。

专业术语来源：《城市轨道交通工程基本术语标准》（GB/T 50833—2012）（*Standard for basic terminology of urban rail transit engineering*）

💡 **Outlines**

_____

_____

_____

Task 1-2: Emergency Response

 Notes

站厅端火灾作业程序

确认火灾报警—组织灭火—确认火灾模式启动—汇报火灾情况—组织疏散—汇报疏散情况—本岗位疏散。

站台门故障处理（单门故障先期处置）

发现故障—现场处理（站务员确认故障门关闭：站务员手指-站务员口呼-列车出

Lesson 10　Emergency Handling 应急处理

Name:_____　Class:_____　Date:_____

清站台后，站务员用LCB钥匙将故障门的LCB转至"自动"位置，取出钥匙-站务员用对讲机汇报值班员-值班员对讲机回复）。

摘编自：2022一带一路暨金砖国家技能发展与技术创新大赛"城市轨道交通运营设计与应急处理"赛项作业程序。

💡 **Outlines**

Task 2. Practice your oral English and Role play

Task 2-1: Scene A

Work in pairs. Suppose you were in the metro and you saw someone is painful...

💡 **Useful expressions/patterns**

Are you all right? You look pale.../ What's the matter with you? You look not well... Answer. Well. I feel dizzy... / My head is aching...

💡 **Outlines**

Task 2-2: Scene B

Work in pairs. Suppose you were an older sister / brother, your younger brother is missing at metro station...

💡 **Useful expressions/patterns**

Oh, no. I couldn't find my little brother... / Oh, My God. My younger brother is missing. I was looking for him everywhere at the metro station and I couldn't...

Please chill out. Miss/sir. Tell us what happened.../ We need you to calm down and tell us where and when did you last see your younger brother...

💡 **Outlines**

Lesson 10　Emergency Handling 应急处理

Name:_____　Class:_____　Date:_____

Self-evaluation

Item	Excellent（90~100分）	Good（80~90分）	Average（60~80分）	Pass（60分）	Fail（<60分）
Be able to use phrases to make conversations					
Take part in pair work and role play					
Improve oral communication skills					
Be able to finish the exercises independently					
Review and preview lessons consciously					

Grade evaluation

Item（项目）	No.（序号）	Criteria（标准）	Score（分数）		
			Self-evaluation（自评）	Peer（学生互评）	Facilitator（指导教师）
General Evaluation（一般评价）（25%）	1	Teamwork, well-assigned jobs（能进行团队合作，能做到分工良好）			
	2	Group discussions for the topic, plan for oral task materials（能进行小组讨论，能准备口语训练任务材料）			
	3	Well-prepared（准备充分）			
	4	Role playing well-performed（角色扮演表现良好）			
	5	Self-confidence（自信）			
Evaluation of Professional Competence（专业能力评价）（25%）	1	Pronunciation, articulation（发音清晰）			
	2	Accuracy, fluency（语言准确、流利）			
	3	Tone of voice, coherent（语调连贯）			
	4	Ability of cross-cultural communication（跨文化交流能力）			
	5	Presentation skills（口语技巧）			

Lesson 10 Emergency Handling 应急处理

Name:_____   Class:_____   Date:_____

Cotinued

Item （项目）	No. （序号）	Criteria （标准）	Score（分数）		
			Self-evaluation （自评）	Peer （学生互评）	Facilitator （指导教师）
Task-based Evaluation （基于任务的评估） （50%）	1	Describe the pictures（描述图片）			
	2	Practice your oral English and Role play（英语口语练习与角色扮演）			
Final Score （合计）		Self-evaluation Score × 20% + Peer Score × 30% + Facilitator Score × 50% （学生自评分数×20%+学生互评分数×30%+教师点评分数×50%）			
Facilitator's Comments （教师评价）					

Task reflection

FACT: What do you get?

FEELING: How do you feel?

FINDING: What do you find?

FUTURE: What shall you do next?

Complementary reading

Defining an emergency

An incident, to be an emergency, conforms to one or more of the following:

- Poses an immediate threat to life, health, property, or environment.
- Has already caused loss of life, health detriment, property damage, or environmental damage.
- Has a high probability of escalating to cause immediate danger to life, health, property, or environment.

Types of emergency

Dangers to life

Medical emergencies including: strokes, cardiac arrest and trauma. To incidents that affect large number of people such as natural disasters include tornadoes, hurricanes, floods, earthquakes, and outbreaks of cholera, Ebola and malaria.

Lesson 10　Emergency Handling 应急处理

Name:_____ Class:_____ Date:_____

Dangers to health

Some emergencies may not be life-threatening immediately, but may have a serious impact on the sustained health and well-being of one or more people (although health emergencies may subsequently escalate to life-threatening).

The causes of health emergencies are usually very similar to those of life-threatening emergencies, including medical emergencies and natural disasters, although the scope of events that can be classified here is far greater than those that threaten life (such as limb amputation, which usually does not lead to death, but if patients want to recover normally, they need immediate intervention). Many life emergencies, such as cardiac arrest, are also health emergencies.

Dangers to the environment

Some emergencies will not immediately endanger life, health or property, but will affect the natural environment and living creatures. Not all institutions think this is a real emergency, but it will have a far-reaching impact on the long-term condition of animals and land. Examples include forest fires and offshore oil spills.

Translation:

(1) dangers to life

(2) strokes

(3) trauma

(4) life-threatening

(5) tornadoes, hurricanes, floods, earthquakes

(6) forest fires

(7) oil spills

📚 Relevant knowledge

Model Worker Spirit

As the chief maintenance expert of Guangzhou Metro Group, over the years, Zhang Chongyang has innovated many maintenance technology precedents in this "inconspicuous" position to fill the gap in the industry, and honors such as "National May Day Labor Medal" "National Knowledge-based Workers Advanced Individual" "Guangdong Provincial Labor Model" and "Southern Guangdong Craftsman" have come to the scene. When the lights are on, the hot and rushing subway line network gradually cools down from the busyness of the day, and it is time for the "behind-

Lesson 10　Emergency Handling 应急处理

Name:_____　Class:_____　Date:_____

the-scenes heroes" of subway maintenance like Zhang Chongyang to play, who adhere to the bottom line of operational safety, from sunset to dawn.

Summary

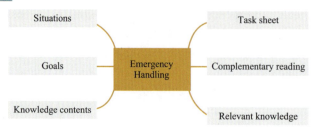

Appendix
附录

Main Role Profile
主要人物角色介绍

Main Role Profile
主要人物角色介绍

	Name : Tina Age : 26 Gender : Female Career : Ticket staff Duties :Passenger service、Ticket service Hobbies : Yoga、shopping
	Name : Allen Age : 21 Gender : Male Career :Metro intern Duties :Passenger service、security check Hobbies : Travelling、delicious foods
	Name : Helen Age : 28 Gender : Female Career : Service agent Duties :Passenger service Hobbies : Painting、films
	Name : Thomas Cook Age : 42 Gender : Male Career :Project manager Hobbies :Gym、plants
	Name : Cherie Age : 33 Gender : Female Career :Tour guide Hobbies : Gym、baking
	Name : Peter Age : 35 Gender : male Career :Resident doctor Hobbies : video games、science-fiction movie、table tennis

References

参考文献

[1] 成应翠, 张燕萍. 精彩世博行[H]. 上海: 华东理工大学出版社, 2010.
[2] 叶清贫. 铁路运输与信号专业英语[M]. 武汉: 华中科技大学出版社, 2008.
[3] 闵丽平. 城市轨道交通专业英语[M]. 北京: 中国铁道出版社, 2006.
[4] 赵娅丽. 交通运输工程专业英语[M]. 上海: 同济大学出版社, 2007.
[5] 陶曙敏, 刘伶利. 轨道交通客运服务实用英语口语[M]. 2版. 北京: 中国铁道出版社, 2009.
[6] 章樊. 马上去旅行: 英语无障碍说走就走[M]. 北京: 中国水利水电出版社, 2015.

学习心得